Dear,

I look forward
to reading your
book when it comes
out. I'm so glad
you are pursuing a
PhD and I look forward
to reading your work.

Aluta continua!

Gladys

Race and the Politics of Knowledge Production

Race and the Politics of Knowledge Production

Diaspora and Black Transnational Scholarship in the United States and Brazil

Edited by Gladys L. Mitchell-Walthour and
Elizabeth Hordge-Freeman

palgrave
macmillan

First published 2016 by
PALGRAVE MACMILLAN

The authors have asserted their rights to be identified as the authors of this
work in accordance with the Copyright, Designs and Patents Act 1988.

Palgrave Macmillan in the UK is an imprint of Macmillan Publishers Limited,
registered in England, company number 785998, of Houndmills, Basingstoke,
Hampshire, RG21 6XS.

Palgrave Macmillan in the US is a division of Nature America, Inc.,
One New York Plaza, Suite 4500, New York, NY 10004-1562.

Palgrave Macmillan is the global academic imprint of the above companies
and has companies and representatives throughout the world.

Hardback ISBN: 978–1–137–55393–5
E-PUB ISBN: 978–1–137–55395–9
E-PDF ISBN: 978–1–137–55394–2
DOI: 10.1057/9781137553942

Distribution in the UK, Europe and the rest of the world is by
Palgrave Macmillan®, a division of Macmillan Publishers Limited,
registered in England, company number 785998, of Houndmills,
Basingstoke, Hampshire RG21 6XS.

Library of Congress Cataloging-in-Publication Data is available from the
Library of Congress.

A catalogue record for the book is available from the British Library.

To my daughter, Truth Ann Walthour; husband, Anthony Walthour; and in memory of my father, David Mitchell Jr.

To McArthur Freeman II and Nathaniel Freeman

Contents

Acknowledgments

We express our deepest gratitude to those in Brazil and the United States who have welcomed transnational researchers into their friendship networks, research teams, and community organizations. We also thank those who contributed to this edited book in order to share their experiences and insights. There are many transnational researchers whose contributions are not included in this book, but they comprise a growing network of black diasporic scholars whose work and personal politics are invested in promoting critical conversations about transnational research. Of the many organizations that have supported the coeditors' transnational research, we would like to recognize the Instituto Cultural Steve Biko, Brazil Cultural, the Federal University of Bahia, and the State University of Feira de Santana. The supportive scholars and activists in Brazil who have been critical to our work are numerous. While we cannot name them all here, we would like to recognize Michel Chagas, Ecyla Saluy Moreira Borges de Jesus, Silvio Humberto, Jucy Silva, Paula Cristina Barreto, Clóvis Oliveira, Edna Araújo, Raquel de Souza, Javier Escudero, Patrícia Burgos, and Nancy Souza e Silva. In the United States, scholars including Dianne Penderhughes, Cynthia Spence, Bernd Reiter, William Darity Jr., and Eduardo Bonilla-Silva have served as mentors who have guided us as we have pursued diasporic research. To our fellow Brazilianists, we are indebted to you for constantly pushing the boundaries of research further and challenging us to be better researchers.

We would like to especially acknowledge Michael Mitchell who was a political scientist and a pioneer in studying Afro-Brazilian racial consciousness and political behavior. Mitchell passed away during the production of this book. He has been instrumental in guiding the work of numerous African American and Afro-Brazilian scholars and activists interested in racial consciousness and social justice.

As coeditors, we would also like to express our appreciation to the person who initially introduced us: Dr. John French, Professor of History at Duke University. As an undergraduate at Duke University, Mitchell-Walthour enrolled in French's popular Afro-Brazilian Culture course, which changed the course of her career. She ultimately continued her studies at the University of Chicago where she earned a PhD focusing on Afro-Brazilians and political behavior in Brazil. In 2007, as a PhD student in Sociology at Duke University, Hordge-Freeman enrolled in French's graduate course, "Research in Global Connections," in which he mentored her as she developed a paper that discussed the positionality and politics of black diasporic scholars in Brazil. French's mentorship of both coeditors and his intellectual contributions to Brazilian scholarship have been invaluable.

Our special thanks to the Humanities Institute at the University of South Florida for providing financial support for the completion of this project. We also wish to acknowledge the assistance of three amazing graduate students at the University of South Florida—Anna Abella (anthropology), Rodrigo Serrão (sociology) and Michelle Angelo Dantas (Latin American Studies)—who worked on the translation and editing. Moreover, Hordge-Freeman would like to recognize the research support provided by the Sociology Department, the Institute for the Study of Latin America and the Caribbean (ISLAC), and the ISLAC Afro-Descendants Working Group at The University of South Florida. She also expresses gratitude to the Fulbright program, Ford Foundation, American Sociological Association, Ruth Landes Memorial Fund and the USF Office of Community Engagement and Partnerships for providing resources for her transnational work in Brazil.

Finally, we thank our families for supporting us through long research trips, writing binges, and our continued interest in Brazil. Specifically, Mitchell-Walthour would like to thank her husband, Anthony Walthour, for always believing in and supporting her. She also thanks Violet Mitchell, Gladys H. Mitchell, Cornelia Smith, Julia Elam, and Davia Lee. Hordge-Freeman would like to acknowledge her husband, McArthur Freeman II, and Patricia Hordge and Jenifer Hordge for being the main engines behind her family support system. Both coeditors recognize their children, Truth Ann Walthour and Nathaniel Freeman, for giving them the best reasons to complete their research and return home.

Introduction: In Pursuit of Du Bois's "Second-Sight" through Diasporic Dialogues

Elizabeth Hordge-Freeman and
Gladys L. Mitchell-Walthour

[T]he Negro is a sort of seventh son, born with a veil, and gifted with second-sight in this American world—a world which yields him no true self-consciousness, but only lets him see himself through the revelation of the other world. It is a peculiar sensation, this double-consciousness, this sense of always looking at one's self through the eyes of others, of measuring one's soul by the tape of a world that looks on in amused contempt and pity. (Du Bois 1903 [1996], 5)

In *Souls of Black Folk*, sociologist W. E. B. Du Bois elaborates on the notion of "double-consciousness," a concept that captures how racial marginalization shapes the perspectives, experiences, and identity of blacks in American society. He characterizes blacks' positionality as one marked by clashing dualities that create "two warring souls, two thoughts, two un-reconciled strivings; two warring ideals in one dark body" (Du Bois 1903 [1996], 5). But he also notes that when this unique positionality is stimulated and directed, it can ultimately give way to heightened awareness and discovery, what he refers to as "second-sight." Du Bois's prolific and pioneering contribution to the social sciences, unmatched by those in his time or ours, is perhaps the best example of the manifestation of this "second-sight" (Morris 2015). His assertions about "double-consciousness" are cited profusely in interdisciplinary and international contexts, but his statements about "second-sight" have not garnered nearly as much attention despite their implications for black researchers' knowledge production.

In this volume, we seek to explore how the racial positionality that informs "second-sight" might be further sharpened and broadened by the experiences of transnational black researchers. Moving beyond the confines of the United States, we extend Du Bois's commentary about "second-sight" to include black (South) Americans from Brazil. While there are differences related to national history, the process and nature of slavery, and the construction of the racial systems in the United States and Brazil (Marx 1998), there is value in exploring the experience of black researchers from both countries (Skidmore 1993; Bonilla-Silva and Dietrich 2009). Foreshadowing the exponential increase in transnational research, Du Bois, who maintained an orientation that was never narrowly confined to the United States, vociferously advocated the establishment of pan-African alliances in order to support a political imperative that involved the "formation of political and cultural movement premised on international solidarity" (Patterson and Kelley 2000, 19). We bring questions of positionality, reflexivity, and embodiment together in this volume to discuss the politics of knowledge production, the challenges of coalition building, and how race, gender, and nationality lead researchers to reconstitute their findings and the way they view themselves. The contributors ultimately illustrate that their positionality as black transnational researchers is dynamically multidimensional and that the identity negotiations that accompany their experiences in the United States and Brazil are the basis on which they can gain insights into the racial systems in these countries and ultimately fulfill Du Bois's notion of "second-sight."

Continuing Black Transatlantic Dialogues

The modern emergence of intellectual discourse that encourages scholars "to think transnationally . . . and move beyond the limits of the nation-state" makes the notion of diaspora particularly appealing because it is a concept that naturally resists political boundaries (Patterson and Kelley 2000, 12). For black scholars, many of whom define themselves as part of a larger international black community, diasporic studies make up the stage on which globalization and transnationalism intersect and interact in order to nurture "imagined communities" (Anderson 2006). What occurs at these particular diasporic junctures, the types of ideas that get transmitted and are later transformed, and the process by which the researcher (re)positions herself or himself as neither insider nor outsider reveal the unexpected outcomes of modern and more accelerated transnational dialogues. Even as debates about interconnectedness continue to contest whether globalization reflects newness or more of the same, we are interested in exploring how self-reflexivity and positionality impact transnational dialogues and knowledge production.

In the twenty-first century, black scholars' interest in diasporic connection represents just one of many moments in the extensive trajectory of "centuries-old black transnationalism" between the diasporas and homelands of the Atlantic perimeter (Matory 2005, 37). But only recently has the term "dialogue" been introduced to refer to the transformative exchange between Africa and the Americas (Matory 1999; Okediji 1999). In contrast to perspectives that have viewed engagement between Africa and the Americas as a unidirectional story of origin, historical encounters between Africans and African descendants reflect a relationship that is "engaged in a transnational dialectic of mutual transformation" (Matory 2005, 71; see also Yelvington 2001). The analytical shift that the notion of dialogue implied was momentous; it broke away from more traditional perspectives that both privileged ideas about one-way cultural transmissions and were limited by narrow comparisons that focused on debating syncretism and survival. Though the population of Africans and African descendants who participate in these Afro-Atlantic dialogues "encompasses speakers of about a half dozen other European languages, a dozen creole languages, and hundreds of African languages," in this volume we focus mainly on exploring the experiences of researchers from the United States and Brazil (Matory 2006, 169).

This edited collection marks a timely contribution to the literature because since David Hellwig's edited book *African American Reflections on Brazil's Racial Paradise* (1992), few books on black intellectuals' experiences in Brazil have been published. Hellwig's anthology outlines the historical trajectory of previous works and intellectual perspectives by black US scholars on Brazil from the early 1800s through 1990. It highlights the significance of first-wave scholars including Lorenz Turner, E. Franklin Frasier, and W. E. B. Du Bois focusing on the ways in which Brazil was framed as a racial paradise. However, it does not fully capture the ideological transition that occurs on the heels of "the most important black social movement of the century," which was organized by the Movimento Negro Unificado (MNU) in the late 1970s (Mitchell 2003, 35). During this period, the racial democracy façade was decried by the majority of US scholars, including Michael Mitchell, Leslie Rout, Anani Dzidzienyo, and J. Michael Turner—many of these scholars have works that appear in Pierre-Michel Fontaine's (1985) earlier edited anthology of *Race, Class, and Power in Brazil*. When considered in the historical context of Brazil's authoritarian regime, these second-generation black US scholars were "inspired by the latter stages of the American civil rights movements and the resurgence of an international outlook" in the United States (Mitchell 2003, 35).

The edited anthologies by both Fontaine (1985) and Hellwig (1992) make a significant contribution to research, but their linear framing of the perspectives of researchers belies the more complex conceptual transformations and identity negotiations that scholars experience. In Fontaine's (1985) published anthology, Carlos Hasenbalg notes that the discourse and academic study of race in Brazil can be classified into three principal stages.[1] In his later book, Hellwig (1992) describes the three ways that the perspectives of black US researchers have changed by organizing his work into a neat, linear procession designated as the Myth Affirmed, the Myth Debated, and the Myth Rejected. However, the presence of black US scholars in Brazil during periods of ideological transitions when ideas about race were being (re)formed meant that understanding how race functioned in Brazil was a multifaceted and complicated endeavor, rather than a linear one. For example, Angela Gilliam's (1992) research piece on Brazil, which is included in Hellwig's anthology, includes a footnote that informs the reader that Gilliam experienced "a permanent ideological shift toward global egalitarianism" and away from black nationalism (173). Therefore, even when the level of interest in Brazil as a research topic of black US scholars has been steady and consistent, the reflections of these scholars might still be characterized by changing perspectives, retractions, and even contradictions. This volume builds on these earlier edited works by critically analyzing the positionality of black researchers with an emphasis on the challenges they face and the strategies they employ in their identity negotiation, strategies that ultimately inform their changing views about race. How racial contradictions manifest themselves in the field and in their daily lives and how scholars manage these personal and intellectual shifts will be addressed.

Black Brazilian and Black US Scholars in Dialogues

Presently, we find ourselves in the midst of what might be called a third wave of scholarship on Brazil, which includes black scholars who emerged in the period between 1995 and 2015. These scholars extended the diverse ideas that surged during the first and second waves to incorporate a focused attention on broader areas of interest falling within interdisciplinary studies and were attuned to intersectionality and feminist theory.[2] Notable in the progression from the first wave to the third wave of scholarship is the growing number of black US scholars writing about Brazil and the increasing number of women contributing to this emerging literature, a fact evident in the overwhelming representation of women as contributors to this volume. Even more exciting than the changing face of black US researchers conducting research in Brazil are the more racially diverse Brazilians entering academia, anchored by legislative victories that have opened a significant number of opportunities for Brazilian students (particularly lower-income and black students) to study at colleges and universities (Htun 2004; Martins et al. 2004).

These new Brazilian researchers in training who are interested in studying race are in good academic company: numerous black Brazilian intellectuals, including Neuza Santos, Milton Santos, André Rebouças, Joel Rufino, Abdias do Nascimento, Lélia Gonzalez, Guerreiro Ramos, and Sueli Carneiro, among many others, have made significant contributions (Figueiredo and Grosfoguel 2007).[3] The presence of this new critical mass of black Brazilian scholars will ultimately challenge the Brazilian academy to shift away from exclusive practices in order to benefit from the theoretical and conceptual contributions that come from the infusion of diverse ideas (Lima 2001; de Carvalho 2007). Ideally, the introduction of new voices into intellectual conversations about race in Brazil will be accompanied by more interdisciplinary perspectives, a deeper interrogation of researcher positionality, new directions in intellectual inquiry, and an attention to scholarly issues that have been historically underdeveloped (such as whiteness). Not only might the presence of emerging black Brazilian scholars push the Brazilian academic establishment to contract and expand, but their insights might also (re)shape how scholars understand race in the United States. As Pereira (2013) clearly demonstrates—in ways that resonate with Matory's ideas about transatlantic exchanges and dialogues—the black movement in Brazil was influenced by and also influenced black US activists. We can expect that a similar exchange may continue to shape research and activism in Brazil and the United States.

In this volume, black US scholars explore their experiences researching race in Brazil and black Brazilian researchers discuss their experiences analyzing race in the United States. Though we focus on only two countries in an expansive African diaspora, we believe the types of misrecognitions, experiences, dilemmas, and possibilities addressed by researchers in these contexts will also resonate with the experiences of other diasporic researchers. Previous reflections by black US researchers in Brazil are instructive in terms of revealing the complex identity negotiations involved in conducting research on race in Brazil. Michael Turner (1985) recounts the dilemma of being a foreign researcher in Brazil, explaining that he could "point to example after example of seemingly blatant discrimination or attitudes indicative of racial prejudice, only to be told by Afro-Brazilians that one had not understood the reality or social context of the situation" (76). Though Turner initially speaks of being a nonracialized researcher, he later mentions that "Black American researchers arriving in Brazil were sometimes treated as survivors of the war," given that popular images of the United States were replete with racial unrest, church bombings, and examples of extreme violence (77). While he does not explore how this special status as a survivor influenced his interactions with Afro-Brazilians, his narrative poses questions about the relationship between positionality and knowledge production.

Addressing positionality more directly, Hanchard (2000), a black Brazilianist and political scientist from the United States, describes his positionality as a moving target that was viewed as an "ally, object of suspicion, confidant, interlocutor, teacher, and student" (167). But the reader is still left wondering how he managed these conflicting identities. Similarly, as Winddance Twine (2000) suggests, in some cases, Afro-Brazilians rejected her as a black researcher and, instead, requested or preferred to speak with her white research partner. Indeed, Twine's experiences challenge the assumptions that "closeness of identity and, in particular, shared racial identity . . . promote effective communication between researcher and subject, and conversely, disparate identity to inhibit it" (Rhodes 1994, 550). While this is true, researchers would benefit greatly from exploring the potential responses to these types of encounters.

Finally, one of the more fundamental questions that diasporic scholars ask themselves is the one posed by Michael Mitchell (2003): "How does a social scientist balance the commitment to scientific detachment with the equally important commitment to praxis and to actually participating in a process of political change?" (36). How to balance one's personal politics and academic career is a question that has continually been asked. Sometimes there are replies, but no concrete answer fits every case.[4] What Mitchell's question does is to forecast the intense debate that has developed in response to the role of US intellectuals in the diffusion of cultural ideas, beliefs, and cognitive orientations about race and color in Brazil. Bourdieu and Wacquant (1999) critique researchers, targeting a particular black US researcher, with accusations that they are seeking to "replace wholesale" the history of Brazil, in order to advance their investment in "facilitating the globalization of American problems" (44). These arguments parallel those made by anxious Brazilian elites who viewed the development of a black movement in Brazil as an "imposition of North American cultural values and racist separatist tendencies on what had been a 'calm and happy'" black Brazilian populace (Turner 1985, 79).

Prominent social scientists have rejected the critique that black researchers simply impose their racial views onto Brazil (French 2000; Hanchard 2003; Sansone 2003; Telles 2003). In a systematic critique, French (2000) methodically identifies how the critiques launched against the work of black US scholars are riddled with "missteps and misjudgments that vitiate the intellectual and political project they [the critics] claim to favor" (114). Caldwell (2007) further problematizes the broader implications of the specific critiques aimed at black transnational researchers that suggest they "lack sufficient objectivity in their research on race" (xx). Black researchers are implicitly assumed to be biased, whereas white researchers are given the privilege of presumed objectivity (Hendrix 2002). This is problematic for a number of reasons, not least because Warren (2000) offers numerous examples of how whites' emotional investment in "sustaining the image of Brazil as a racially meritocratic society" often leads them to discount racism even when it is consequential for their analysis (145). To be certain, an awareness of the role that the United States plays in shaping international discourses and reinforcing hierarchies of knowledge is essential, and when substantiated with compelling evidence, these asymmetries can begin to be addressed (Thomas 2007). The inclusion of insights from black Brazilian researchers takes on even greater importance in light of this ongoing debate.

Racial Embodiment and Knowledge Production

Black diasporic researchers have written about their positionality in transnational settings, and often about their ability to conduct research in diasporic countries relatively undetected. This ability to blend in with the diasporic communities in which they study is sometimes connected to their racially marked bodies, but may also be connected to certain cultural behaviors, including linguistic fluency, manner of dress, and attitudes. While racial similarity can certainly be a point of advantage, black researchers' positionality and body politics in the field are complicated because, as Twine (2000) suggests, racial insiderness "may simply create a different set of pluses and minuses," rather than eliminate them (13). What has been much less studied is *how* researchers navigate their racial subjectivities in ways that lead to intellectual and personal "transformation through embodiment" (Okely 2007, 77). For black researchers there are often multiple levels of evaluation at play, including their racialization, based on characteristics deemed significant in their country of origin and those which are deemed socially significant in their host country. Their negotiations of these conflicting systems of racial categorization and hierarchy emerge through their embodied experiences and may ultimately shape their formation of research questions, selection of theory orientation, and interpretation of the analysis.

Speaking of the importance of how embodiment and race impact women, Motapanyane (2010) argues that "transnational research ethics have yet to adequately grapple with the complicated dynamics of research on regions of the global south that is conducted from first-world universities by women of colour who are themselves racialized and othered in their academies" (100). Rather than suggesting that women and people of color are beyond reproach in terms of questions of power and equity, she suggests that they, too, may engage in "acts of othering and exploitation" that reflect and reproduce their internalization of a particular type of engagement with third-world women

(100). To that end, this volume explores the extent to which black researchers recognize and negotiate the relative privileges that they may enjoy based on race, gender, and nationality. Further explaining that racial appearance is not a proxy for political solidarity, Subreenduth and Rhee (2009) argue that "[t]he color of body whether it is yellow, brown, black, or red, does not guarantee innocence, resistant points, or authenticity as a voice for anti-oppression. In fact, different shades of a color can be a source of discrimination and oppression within one's own community" (338). Though a difficult issue to broach, our contributors confront how power and privilege shape their experiences, direct their research trajectory, and impact their notions of race.

Globalization and Racial Contestation

Researchers' embodied experiences serve as sources of scholarly insight, but may also have a broader impact on one's perceptions and investment in notions of global blackness. Black Brazilian feminist Lélia Gonzalez coined the term "Amefricanidade" as a way to articulate her "transatlantic vision of black feminism," which she hoped might include constructing a transnational network of black women from across the diaspora who could work together to promote racial and gender equality (Williams 2014, 9). As Gonzales's vision suggests, these networks and alliances must be constructed, because their existence is not inevitable. In the same way, while black researchers may be uniquely positioned to fulfill Du Bois's notion of "second-sight," doing so is "neither inherent nor innate," but rather it is achieved by several factors including lived experience, targeted efforts, and training (Holt 1990, 306). In this volume, four Brazilian researchers highlight the critical role of black American culture (as experienced through music, film, and television shows) in shaping their developing racial identity, notions of blackness, and interest in participating in racial resistance in Brazil.

There is considerable evidence that US repertoires of resistance including the "black is beautiful" slogan, cultural innovation, and political organization have been integrated into other transnational and antiracism contexts (Pinho 2010; Cumberbatch 2009). While some contributors in this volume might view this influence as beneficial, others remain vigilant of what Thomas (2007) refers to as "the inequities and hegemonies within diaspora," which describes the unequal cultural influence of the United States on other diasporic societies (56). The geographic expansiveness of diasporas suggests that while the political project of diasporic connections and national similarity could forge a powerful political alliance, the practical requirements necessary to unify such massive and diverse populations are tremendous, and must be continually critiqued and (re)evaluated. The scholars in this volume discuss their movements through shifting racial contexts, their subjection to various forms of embodied racialization, and their exposure to transnational influences in order to reveal how their positionality is constructed and negotiated in ways that advance their work in the United States and Brazil.

Organization of the Book

The book is organized into three main parts. Part I provides a broad overview of the complex roles of academic institutions in shaping the politics of knowledge production. This section begins with Chapter 1 by Kia Lilly Caldwell in which she uses a feminist approach to explore the tenuous position of black Brazilian women within women's studies. Caldwell draws on her experiences as a Brazilianist researcher and professor in the field of African American/ African diaspora studies to highlight the invisibility of black Brazilian women in the field of black women's studies in the United States. She ultimately argues that their absence in academic scholarship is a reflection of their broader political, social, and economic marginalization. In Chapter 2, David Covin discusses the Race and Democracy Project, which he argues is the most productive collaboration created to date that has linked African-descended scholars from the United States and Brazil. He traces the evolution of the collective as a transnational project and substantiates his argument about its effectiveness by outlining the impressive cadre of Afro-Brazilian participants who have been trained in the program, which now includes several doctorate degree holders and a current member of the Brazilian federal government. In Chapter 3, Elizabeth Hordge-Freeman explores how her positionality has shaped her conceptualization of diasporic engagement and has impacted the development of a summer program in Salvador, Bahia, Brazil. She explores how she negotiated both challenges and privileges involved in the program's development, including questions of cultural voyeurism, reciprocity, and critical global citizenship. She ends the chapter by critically reflecting on the program's limitations and outlining how diasporic researchers can use institutional structures and local community partnerships in global contexts to foster the type of collaborations needed to produce research that is responsive to black diasporic communities. In Chapter 4, Mojana Vargas discusses the shifting ways that her location (in the United States, Brazil, and Portugal) has shaped how her blackness and racial features (especially her hair) have been policed, albeit in different and ironic ways. She ends her chapter with an analysis of how she has handled institutionalized academic practices that undermine the well-being and successful integration of black students through their reinforcement of dominant hierarchies of status, aesthetics, and culture.

In Part II, contributors discuss the field as the site of their research, and also the place where they negotiate their own racial/gender/sexual identities in ways that lead to new substantive and conceptual discoveries. In Chapter 5, Tiffany D. Joseph discusses how her racialization (as nonblack) in Governador Valadares leads her to reflect on and shift her own conceptualizations of race in the United States. She provides examples of how her theoretical concept—"transnational racial optic"—emerged from the experience and concludes by exploring the implications of using one's subjectivity to better understand the nuances of social constructions of race. In Chapter 6, Jaira J. Harrington uses her experiences in Brazil in a student homestay program, a language school student, and as a student conducting fieldwork to illustrate how the "aesthetics of power" leads to constant confusion and misidentification about her identity when she meets others. She views her misidentification as a domestic worker as providing both inspiration and insight to her research on domestic work in Brazil. In Chapter 7, Reighan Gillam uses a conceptual idea of "African diaspora looking relations" to discuss the ways that her body and racial features were (mis)read and contested in order to reveal the constructedness and unpredictability of race. She illustrates that the transnational gaze is bidirectional, engaged by both US blacks and black Brazilians, in ways that reflect the particularities of racialization processes. In Chapter 8, Gladys L. Mitchell-Walthour discusses how as a dark-skinned black woman and US researcher in Brazil she experienced racism in ways that allowed her to more deeply and critically analyze and interpret blackness beyond its use as a political identity. The insights that she gains from these subjective experiences inform her conceptualization of taken-for-granted terminology often used in social science research. In Chapter 9, Chinyere Osuji discusses the tensions surrounding race, gender, and sexuality during her fieldwork which emerged during her interviews with interracial couples in Los Angeles and Rio de Janeiro. She examines how the "angry black women" stereotype and notions of black women as "leftovers" in dating and marriage markets were ever-present ideas in her interactions with her interviewees. Her chapter explores how black women—both inside and outside of these relationships—give meaning to black–white couples.

In Part III, Afro-Brazilian scholars and graduate students who are currently in the United States discuss their experiences negotiating race and ethnicity in the United States. There are several features that make these chapters unique. They reflect experiences of Afro-Brazilians in various cities in the United States, including Philadelphia, Los Angeles, Austin, and Atlanta. Moreover, the contributors attribute their interest in race relations to their early inspiration from and exposure to African American culture while in Brazil. In Chapter 10, Daniela F. Gomes da Silva describes early exposure to black culture in the United States through music in Brazil as critical to her racial identification. Charting her interactions and observations in Atlanta, Georgia, she discusses how idealistic notions of racial solidarities and political activism were belied by the actual experiences that she encountered in the United States. In Chapter 11, Lúcio Oliveira reveals that he was similarly drawn to black American culture through music but describes having no delusions that the United States would be radically different from Brazil. He begins his chapter with an encounter with the police in Los Angeles in order to explore the familiar ways that black male bodies are racialized in Brazil and the United States. He provides examples of racialized encounters from childhood (at school and at home) that lead to his interest in the United States and notes how his inclusion in particular academic networks involving both black Brazilian and US scholars has been essential to fostering his critique of racial democracy and whiteness in Brazilian society. Finally, in Chapter 12, Gabriela Watson Aurazo discusses the way that growing up in Brazil exposed her to popular black sitcoms, which fostered a source of racial pride and affirmation, and led her to pursue an MA in film studies at Temple University. She describes the process of discovering that the images of racial radicalism were only one element of African American culture, and arguably not the most salient.

Notes

1. The first stage is closely associated with the work of Gilberto Freyre, who originally coined the term "racial democracy" to describe what he believed was the lack of racism and racial discrimination in Brazil. The second phase occurred during the 1940s and 1950s, a period in which Brazilian scholars from the northeast region of Brazil predominated with theories that attributed enduring inequality to class-related inequalities. The third school of thought occurred during the 1950s and 1960s, and the ideas that emerged during this period completely countered the first two phases because they focused explicitly on the role of racism and racial discrimination in the life chances of Brazilians (Fontaine 1985).
2. Among some of these more prominent scholars are Michael Hanchard, Kim Butler, Kia Lilly Caldwell, France Winddance Twine, Michael Mitchell, Michael Turner, Angela Gilliam, Melissa Nobles, Keisha Khan-Perry, Erica Williams, Christen Smith, and Tianna Paschal.
3. Figueiredo and Grosfoguel (2007) wrote a compelling article about the invisibility of black academics. In addition to the names included in this chapter, they refer to several other scholars whose work has paved the way for critical discussions of race in Brazil.
4. Reiter and Oslender (2015) recently edited a book on academic scholarship and activism with participation from several Brazilianists who reflect on how they balance activism in the scope of their research, including contributions by Fernando Conceição, Keisha-Khan Perry, Christen Smith, and Elizabeth Hordge-Freeman.

PART I

Institutions as Gate-Keepers and Game-Changers

Black Women's Studies in the United States and Brazil: The Transnational Politics of Knowledge Production

Kia Lilly Caldwell

I was first drawn to Brazil during my sophomore year in college as a result of an eye-opening presentation made by Joselina da Silva, a black Brazilian activist and scholar. She spoke about the political work black women were doing in Rio de Janeiro. Until that point, I had little to no knowledge of African-descendant communities in Brazil and my curiosity was piqued. Over the past 20 years, my personal and professional endeavors have been focused on trying to better understand black women's experiences in Brazil and also contributing to efforts to promote racial and gender justice, especially through my research and scholarly projects. As someone who conducts research and teaches courses on black communities in Brazil, other areas of Latin America, and the United States, as well as courses on gender, I often think comparatively about knowledge production in and about these different geographic spaces and ways to challenge the invisibility of black women, particularly in academic research. In this chapter, I reflect on the development of the field of black women's studies in the United States and some of the challenges to increasing the production of scholarship focused on black women within the academy in Brazil. I also examine some of the problems associated with incorporating black Brazilian women's experiences and writings into women's studies in Brazil and black women's studies in the United States.

My research on issues of race and gender in Brazil, since the early 1990s, has been both professionally and personally rewarding. However, as I have noted in previous work (Caldwell 2007), forming alliances—both academic and political—across different African diasporic communities is not automatic or without challenges. Over the years, I have discovered the importance of having key allies and colleagues in Brazil, in order to establish relationships with other individuals and organizations for my research, as well as to solidify my professional credentials and reputation. Being a black woman from the United States has likely facilitated some of these relationships, but it has also been important for me to demonstrate continued commitment and solidarity to feminist and antiracist struggles in Brazil over long periods of time. While as an individual researcher, it has been rewarding to contribute to scholarship on race and gender in Brazil, I also realize that a collective effort is needed, both inside and outside of Brazil, in order to move work in this area forward in a sustained manner. It is in this spirit that I share my reflections on the opportunities and challenges associated with the development of black women's studies in Brazil.

Creating the Field for Black Women's Studies in the United States

During the nineteenth century, there was an upsurge in social and political activism among black women in the United States and an increase in the number of literary works, political essays, and journalism produced by them.[1] Publications and public speeches by black women highlighted the role of race and gender in shaping their experiences during slavery and in post-abolition US society. At the end of the 1970s and the beginning of the 1980s, there was a resurgence in written work focusing on black women in the United States. During this time period, black women began to develop a critique of the second-wave feminism of the 1960s and early 1970s and also of the civil rights and black power movements. These political critiques, like the work of black women during the nineteenth century, emphasized the importance of race and gender.

The writings of US black women during the 1970s and 1980s became the basis of a newly forming field known as black women's studies. Pioneering texts during this period include *The Black Woman: An Anthology* (1970) edited by Toni Cade Bambara and Elenora Traylor and *All the Women Are White, All the Blacks Are Men, But Some of Us Are Brave: Black Women's Studies* (1982) edited by Gloria T. Hull, Patricia Bell Scott, and Barbara Smith. The latter is an important marker in efforts to define and institutionalize the field of black women's studies in the United States. This collection of essays written by black women delineates themes that are relevant to their experiences and emphasizes the necessity of developing a field that would focus on issues relevant to black women in North American universities. In the introduction to *But Some of Us Are Brave*, two of the book's editors, Hull and Smith (1982), speak of the significance of research and writing about black women:

> Merely to use the term "Black women's studies" is an act charged with political significance. At the very least, the combining of these words to name a discipline means taking the stance that Black women exist—and exist positively—a stance that is in direct opposition to most of what passes for culture and thought on the North American continent. To use the term and to act on it in a white-male world is an act of political courage. (xvii)

Hull and Smith went on to note the following:

> Like any politically disenfranchised group, Black women could not exist consciously until we began to name ourselves. The growth of Black women's studies is an essential aspect of that process of naming. The very fact that Black women's studies describes something that is really happening, a burgeoning field of study, indicates that there are political changes afoot which have made possible that growth. To examine the politics of Black women's studies means to consider not only what it is, but why it is and what it can be. (xvii)

Hull and Smith call for an expanded notion of politics that includes "any situation/relationship of differential power between groups or individuals" (xvii). Their essay also highlights four themes they consider to be important in the politics of black women's studies:

> (1) the general political situation of Afro-American women and the bearing this has had upon the implementation of Black women's studies; (2) the relationship of Black women's studies to Black feminist politics and the Black feminist movement; (3) the necessity for Black women's studies to be feminist, radical, and analytical; and (4) the need for teachers of Black women's studies to be aware of our problematic political positions in the academy and of the potentially antagonistic conditions under which we must work. (xvii)

Hull and Smith (1982) develop the fourth theme cited above later in the essay by commenting that the working conditions within the "white-male academy" exist within a structure that is not only elitist and racist, but also misogynist (xxiv). In addition, they point to the fact that black women are part of two groups that have been considered "congenitally inferior in intellect," that is, blacks and women (xxiv). Hull and Smith argue that suppositions regarding the intelligence of black women threaten their credibility and legitimacy as intellectuals within the North American academy.

Hull and Smith point to the necessity of rejecting modes of thought derived from "white-male Western thought" (xxiv). They also speak about the importance of developing methods to minimize the isolation of black women within the "white-masculine academy" and of constructing the types of support networks that black women have always formed to help one another to survive. Finally, Hull and Smith argue for the necessity of creating spaces that belong to black women intellectuals, such as conferences, journals, and institutions, where black women can feel at home and gain respect for the "amazing depth of perception" that accompanies their identities (xxv).

During the 1980s and 1990s, the majority of studies about black women in the United States were in the fields of history and literature. There was a marked increase in the number of books and studies published about black women during the 1980s. Black women scholars such as Darlene Clark Hine, Paula Giddings, Nell Painter, Sharon Harley, Rosalyn Terborg-Penn, Jacqueline Jones, and Deborah Gray White were part of a new generation of researchers who focused on black women's experiences, struggles, and forms of resistance in slave society and in the postslavery era.[2] Until the 1980s, black women were either invisible in history texts or were stereotyped as "mammies" or "jezebels." In their research, black women historians of the 1980s and 1990s reclaimed the contributions of black women to the social and economic development of the United States and valorized their central role in the development and survival of the black community, from slavery to the present. These studies also contributed to making this new focus on black women's history an important and viable scholarly endeavor.

During the 1980s, the work of black feminist literary scholars, such as Barbara Christian, Nelly McKay, and Hazel Carby, on black women's literary traditions in the United States,[3] had a major impact in fields such as English and African American studies and played a crucial role in breaking white male hegemony in the field of literature, particularly since the "canon" of great works historically excluded women and people of color.

It was also during this period that writings by black women were included in courses at the undergraduate and graduate levels and publishers began to republish literature written by black women during the nineteenth century and early twentieth century that had largely been forgotten.[4] Literary work by black women that was published or republished included novels and poetry by Toni Morrison, Alice Walker, Audre Lorde, Maya Angelou, and Toni Cade Bambara. Beginning in the 1970s, there was a marked increase in feminist theory production by black women such as Angela Davis, bell hooks, Audre Lorde, and Patricia Hill Collins.[5] Black feminist theorists were instrumental in deepening the analysis and understanding of black women's social, economic, and political marginalization in the US context. Hence, the contributions of black women intellectuals, inside and outside of the academy, during the 1980s and 1990s were fundamental to the establishment of black women's studies during this period.

Intellectual Production by and about Black Women in Brazil

Intellectual literature by and about black women in Brazil has often taken place outside of formal academic contexts and much of it was produced during the late twentieth century.[6] As scholars such as Paulina Alberto (2011) have noted, during the early to mid-twentieth century, black men dominated intellectual spaces, such as black newspapers, in Brazil. As a result, there is no well-developed written record of black women's intellectual contributions during this period. However, an important intellectual tradition by black Brazilian women began to emerge during the late 1970s and 1980s with the writings of Beatriz Nascimento, Lélia Gonzalez, Sueli Carneiro, Thereza Santos, Edna Roland, Luiza Bairros, and Fátima Oliveira.[7] The activism of these black women was instrumental in the construction of black feminist theory in Brazil.

Black women's nongovernmental organizations (NGOs) were also important producers of knowledge about black women, although these publications did not follow dominant practices of academic legitimization that are especially important in the United States, such as peer review or publication by a scholarly press. Instead, much of this work was published by black women's organizations, likely due to the lack of more formal publishing opportunities in Brazil as well as cost. These publications typically sought to reach a broader public than most scholarly publications did. Organizations such as Geledés, in São Paulo, and Criola, in Rio de Janeiro, were instrumental in publishing statements, special reports, and books that focused on black women and documented the development of black women's movement. Individual black women also produced important works, some being both scholars and activists, although they did not typically teach in university settings. However, since much of this work was not included in academic publications, it might be devalued and viewed as activist, rather than as scholarly, by some observers. In addition, most of this work does not seem to have become a central part of women's studies, research, or curricula in Brazil.

When I began researching black women's experiences in Brazil during the early 1990s, I encountered major roadblocks in my efforts to study race and reproductive health. I traveled to Brazil for the first time in December 1994 and met with members of the health program at the black women's NGO Geledés, in an effort to learn more about reproductive health issues affecting black women. During this trip, I had an opportunity to meet Edna

Roland, who was then the director of the health program at Geledés, and I also utilized the documentation center at Geledés, which contained important material on black women and their movement. While in São Paulo in 1994, I met Fátima Oliveira, a leading black feminist health researcher and activist, Elza Berqúo, a leading demographer, and other researchers affiliated with the Black Women's Health Program at the Brazilian Center for Analysis and Planning (CEBRAP), which Berqúo coordinated. Although interesting and important research and activism related to black women's health was ongoing in places such as São Paulo during this time, it was difficult for me to undertake research on race and reproductive health, particularly as a foreign researcher, since health data by race was rarely collected and basic epidemiological and health information about the black population was scarce.

It was far less likely that funding agencies would support work in this area during that time, as compared to today, since there was still a widespread belief that race and racism were not important issues in Brazil. This belief had a decisive impact on what types of questions and issues were considered to be legitimate within academic circles and undermined efforts to focus on the experiences of black women. Although my research as a doctoral student and my subsequent book (Caldwell 2007) did not focus on black women's reproductive health, these areas continued to interest me and I have returned to the issue of health in my more recent work. I was able to conduct this research largely because of changes with respect to data collection, research, and public policy that have taken place in Brazil, especially after the World Conference against Racism in 2001 and the creation of government entities such as the Secretariat for the Promotion of Racial Equality Policies (SEPPIR).[8] It should also be noted that these changes have largely been the result of the efforts of black women activists and activists in the black movement, which demonstrates the reciprocal relationship that exists between political struggles inside and outside of the academy.

Prospects for Developing the Field of Black Women's Studies in Brazil

In a special dossier on black women that was published in the Brazilian feminist journal *Estudos Feministas* in 1995, longtime black feminist Matilde Ribeiro wrote about the difficulties she faced in organizing the initial project, which was intended to open

> a space for black women authors that were conducting specific research or developing theoretical formulations about questions of gender and race, political

> participation, or as members of the black movement, feminist and black wom-
> en's movements, academia, public institutions, to have contributions that would
> give a picture of black women and their struggles in the country. (434)

Ribeiro notes that, after almost one year's work, this format did not work. She also mentions the small number of texts that was received for this project. Her comments about the possible causes of what she calls a "fragile response" to the requests to submit articles for the dossier are illuminating for analysis in this field. She attributes the weak response to three causes: (1) the fact that black women's critiques of the feminist and black movements had not been systematized; (2) the distance that existed between academic spaces and social movements, especially with respect to racial questions, and the fact that very few black women were found in academic spaces; and (3) the distance that existed between the practices and theoretical formulations of the feminist movement and the reality of black women.

In considering the advances and continuing challenges in the efforts to establish the field of black women's studies in Brazil, it is useful to think about the continued relevance of Matilde Ribeiro's observations. Many of the issues she identified nearly 20 years ago continue to thwart efforts to amplify the presence of black women and their experiences within the Brazilian academy. As was true in the United States, the presence of black women within academic spaces in Brazil is one of the most important ways to promote their voices, identities, and experiences within university curricula and research agendas of students, faculty, and other researchers. In a 2010 study of the number of black women with doctorates in Brazil, Joselina da Silva found that between 1991 and 2005, out of a total of 63,234 individuals, both male and female, holding doctorates only 251 were black women. That amounts to 0.39 percent. The abysmally low representation of black women at the doctoral level translates into their near invisibility within academic institutions, thus making it more difficult to legitimize black women's studies as a valuable area of scholarly inquiry. In addition, in order for black women's studies to move from the margins and become more "mainstream" its importance will need to be recognized by scholars who are not black women, particularly white male and female scholars.

The fact that Brazil's leading feminist journal, *Revista Estudos Feministas*, frequently publishes work on black women is a move in the right direction; however, there continues to be a paucity of scholarly publications, especially books, focusing on black women in Brazil. To date, no book has been published in Brazil that would be the equivalent of *But Some of Us Are Brave*. While I realize that having this type of text is not a requirement for establishing black women's studies in Brazil, *But Some of Us Are Brave* might serve as a useful model of the type of intellectual manifesto that might be needed to define and move the field forward. In addition, although the ideology of racial democracy has been challenged in many Brazilian academic circles, a good amount of feminist research continues to omit discussion of race and potential differences among women of different racial backgrounds. Such practices perpetuate the invisibility of black women within academic scholarship and undermine efforts to develop a significant corpus of literature on black women's experiences in Brazil.

It is important to recognize that the relative invisibility of black women, both physically and as subjects of scholarly inquiry within the Brazilian academy, reflects their broader social, economic, and political marginalization. This invisibility also maintains a status quo, particularly with regard to black women being continually viewed as *empregadas domésticas* (domestic workers) and social subordinates within Brazilian society (Caldwell 2007). These dominant social images undermine efforts to see black women as thinking beings, intellectuals, or a group worthy of study and critical reflection. As Hull and Smith poignantly observed over 30 years ago in the US context, "[h]ow can someone who looks like my maid (or my fantasy of my maid) teach me anything?" (1982, xxiv). This continues to be relevant in Brazil, given the high number of black women who work in domestic service and larger asymmetries of race, gender, and class in the country. The development of affirmative action policies in Brazil since the early 2000s reflects increasing societal and governmental acknowledgment of the need for proactive measures to address racial inequalities; however, since these policies are relatively recent, it is difficult to gauge their long-term impact in improving educational and employment opportunities for black women.

Several years ago I considered some of the reasons why scholars in the field of women's studies in Brazil seemed to shy away from research and analysis focusing on black women (Caldwell 2000). At that time, I was troubled by the lack of engagement with questions of race. Although critiques of universalist views of women had become commonplace in countries such as the United States and Great Britain, they did not seem to have taken hold in Brazil. In my view, this lack of engagement was largely due to the continued hegemony of the ideology of racial democracy during the late 1990s and early 2000s, as well as the absence of discussions on white racial privilege within women's studies and other academic fields in Brazil. Even today, more than a decade later, the amount of scholarship focusing on the experiences of black women in Brazil is not proportional to their representation in the population, given that the Afro-descendant population was estimated to be slightly over 50 percent of the total population by the 2010 census. A good deal of recent research on black women in Brazil has also been done by scholars from the United States and Canada, which raises interesting and important questions about black women's studies in a diasporic context.

Strengthening Links across Borders

While a detailed discussion of the transnational dynamics and politics of black feminist scholarship is beyond the scope of this chapter, it is important to recognize how national and regional power asymmetries shape the flow of people, ideas, and texts across real and virtual borders. As a scholar whose work engages black feminist activism and scholarship in Brazil, the United States, and other areas of the African diaspora, I have often been surprised by the lack of discussion of key black feminist activists and texts from non-US sites among black feminists in the United States. This observation is particularly relevant to Afro-Latin American feminists and texts and seems to be magnified in the case of Brazil. When I first began my research in Brazil over 20 years ago, I attributed the US-centric orientation of black feminist scholarship in the United States to language barriers, especially for researchers in the United States who were not proficient in reading or speaking Spanish or Portuguese. However, while recognizing the role that language barriers play in obstructing black feminist exchanges across geographical and linguistic borders, it is also important to examine how larger power differentials shape which people, ideas, and texts become central concerns of scholars in the US academy, including black feminists.

While, as previously stated, women's studies in countries such as Brazil have often failed to adequately incorporate the experiences of black women, it is also important to call attention to the relative invisibility of non-US black women in US black feminist scholarship. Over the years, this invisibility has been especially striking to me, given that black feminist activists and scholars in Brazil are usually quite familiar with black feminism in the United States and commonly cite scholars such as Patricia Hill Collins, Angela Davis, Kimberlé Crenshaw, and bell hooks. Several of these US scholars have also given lectures in Brazil and those such as Kimberlé Crenshaw and Angela Davis have frequently traveled there and collaborated with Brazilian feminist activists and scholars. Work by US black feminists has also been translated into Portuguese and published in Brazil; however, the reverse has rarely occurred. While it would be an overgeneralization to say that black Brazilian women's experiences are completely absent from black feminist work in the United States or that there is no engagement between black feminists from the two countries, I believe it is important to highlight the continuing challenges in making the work and experiences of black Brazilian women more visible in the US academy. The pioneering work of anthropologist Angela Gilliam helped to build bridges between black feminist scholarship in the United States and Brazil during the 1990s (Gilliam 1998, 2001; Gilliam and Gilliam 1999). Gilliam was also one of the only US-based black feminist scholars who worked in Brazil at the time. Recent scholarship by several US-based researchers has continued to link black feminist perspectives in both countries and has provided important insights into the impact of violence, urbanization policies, and sex tourism on black women in Brazil (Perry 2013; Smith 2013; Williams 2013). It is also interesting to note that much of this recent work has been done by anthropologists. It is likely that the tradition of traveling to non-US communities and conducting research within the field of cultural anthropology may lend itself to exploring transnational links among black women in ways that are not as common in other fields, particularly those that are more text-centered.

Recent feminist scholarship examines the transnational politics of feminist translation by focusing on how texts and ideas travel and are translated across the borders of the Global North and South, particularly within and across the Americas (Alvarez et al. 2014; Costa and Alvarez 2014). Claudia de Lima Costa and Sônia Alvarez (2014) have highlighted the importance of translation for feminists to cross geopolitical and theoretical borders. As they argue, translation is "politically and theoretically indispensible to forging feminist, prosocial justice and antiracist, postcolonial, and anti-imperial political alliances and epistemologies" (557–558). Costa and Alvarez also call attention to the role of "translocal feminist translational politics" in bridging differences, as well as minimizing mistranslations and misunderstandings among women who may share the same languages and cultures (558). Feminist translation, thus, may be seen as a way of encouraging conversations across boundaries of culture and location that have historically been used to divide women. These observations highlight how a translocal and transnational politics of translation across black feminist spaces is an important first step toward enhancing black feminist dialogue across borders.

While many of the issues outlined above are directly related to black feminist scholarship, they are also deeply intertwined with larger questions of how categories of knowledge are defined in the US academy. Traditional divisions between ethnic studies and area studies can serve to perpetuate the invisibility of Afro-Latin American communities in courses, scholarly associations, and publications. However, as the increased focus on Afro-Latin Americans among US scholars over the past decade has shown, such challenges are not insurmountable. In addition, as fields such as African American studies continue to incorporate a diasporic perspective that pays greater attention to experiences of black communities outside the United States, linking black women's experiences and black women's studies across borders is likely to become an easier and more commonplace practice.

Notes

1. Black women in the United States such as Sojourner Truth, Maria W. Stewart, Anna Julia Cooper, and Ida B. Wells-Barnett played a fundamental role in developing black feminist critiques during the nineteenth century. Harriet Jacobs published the first narrative by a former female slave in the United States during the nineteenth century. See Guy-Sheftall (1995) and Waters and Conaway (2007) for discussions of black women's writing and activism in the United States during the nineteenth century. A similar tradition has not been documented for Brazil; however, that does not mean one did not exist.
2. Relevant works include Hine (1990), Hine and King (1995), Giddings (1984), Harley and Terborg-Penn (1987), and White (1985).
3. See, for example, Christian (1980, 1985), McKay (1988), and Carby (1987).
4. An example of this is Harriet Jacobs's narrative, *Incidents in the Life of a Slave Girl*, which was published in 1861 and was printed only once in the nineteenth century. It was recovered by women's history scholars in the early 1970s and began to be reprinted in 1987.
5. Examples include Davis (1981, 1989), hooks (1981, 1984), Collins (1990), and Lorde (1984).
6. I want to thank Ângela Figueiredo for pointing out the longer tradition of black women's writing that exists in the United States than in Brazil. This has made it more difficult to create a corpus of scholarly texts focusing on black women.
7. See, for example, Bairros (1991), Carneiro (1990, 1995), Carneiro and Santos (1985), Gonzalez (1982, 1988), Oliveira et al. (1995), and Ratts (2007). Black feminists such as Fátima Oliveira and Sueli Carneiro have also published extensively in newspapers, as well as in academic publications.
8. SEPPIR was created by President Luiz Inácio Lula da Silva in March 2003. It is a cabinet-level federal ministry. Matilde Ribeiro, a longtime black feminist, served as the first minister for SEPPIR.

CHAPTER 2

The Genesis of the Race and Democracy in the Americas Project: The Project and Beyond

David Covin

In 1986 K. C. Morrison and David Covin attended a conference of the Association of Caribbean Studies in Salvador, Bahia, Brazil. For both, it was their first trip to Brazil. Although each was active in the National Conference of Black Political Scientists (NCOBPS), they had not met until earlier in the year when they presented papers on the same panels at both the National Council for Black Studies (NCBS) and the NCOBPS annual conferences. By the time they arrived in Salvador, they were fast friends. Neither spoke, read, wrote, or understood Portuguese. Nevertheless, within two days each found himself in the close company of Brazilians. Morrison was befriended by two young Afro-Brazilians, Antônio Rosário de Lima and Júlio Romário da Silva. De Lima was fluent in English and was self-taught in the language by listening to R & B records from the United States (his favorite artist was Sam Cooke), by listening to English language radio broadcasts, and by practicing on black tourists from the United States. Covin was befriended by Laís Morgan, a white Brazilian married to a black college professor from the United States. Over the course of several days both men were introduced to people and places they never would have had access to on their own. For several days de Lima and da Silva took the two under their wing and escorted them to markets, neighborhoods, and gatherings of Afro-Brazilians in the mornings and afternoons, leading them back to their conference hotel, the Hotel da Bahia Tropical, in the evenings. There the four men shared *batatas fritas*, *sanduiches de queijo*, and *caipirinhas*. The situation amused the two Brazilians because—as they explained—without the Association of Caribbean

Studies Conference at the hotel, they would have been denied entry. Yet here they were, eating, drinking, and enjoying themselves as if it were something they did every day.

As a parting gift, Morgan gave Covin a copy of *Lugar de Negro*, written by Lélia Gonzalez and Carlos Hasenbalg (1982). "Read it," she said. "It will teach you the truth about what's going on in Brazil right now." She knew perfectly well he could not read a word of the language, but she wanted to impress him with the book's significance. Morrison and Covin left Brazil, agreeing that they had been introduced to a world they had never known, that it needed to be exposed to black people in the United States, particularly the scholarly public, and that they had to return to the country to learn as much as they could.

At the time, Morrison had an ongoing program of research in West Africa, and could not immediately commit to immerse himself into the study of Afro-Brazilians. He vowed, though, that this was a subject to which he would return, as soon as it was feasible. In contrast, Covin, already deeply immersed in the study of black social movements, had learned enough during their visit to know that a social movement was going on in Brazil. If he studied Afro-Brazilian social movements, he could enrich and amplify his study of social movements writ large. Besides, the indecipherable *Lugar de Negro* was burning a hole in his pocket. To top everything off, when he returned to his campus, in his office, he found waiting a copy of *Race, Class, and Power in Brazil* (Fontaine 1985). He read it and became completely hooked. He created an agenda which included reading everything he could about Afro-Brazilians; teaching himself to read Portuguese from tapes, vinyl records, textbooks; and painstakingly translating, with a pile of Portuguese–English dictionaries beside him, *Lugar de Negro*, word by word.

He committed himself to exposing the Afro-Brazilian world, including its dynamic social movements, to as large a scholarly audience as possible. The question was how to do this without knowing Portuguese. So he began small, with book reviews. The first of these was a review on *Race, Class, and Power*, which appeared in *The Western Journal of Black Studies* (Fall 1987). He realized, however, that more book reviews would need a framing contextual background. That recognition led to two articles on the Unified Black Movement (MNU): one in the *Journal of Third World Studies*, "Ten Years of the MNU: 1978–1988" (Covin 1990a); and the other in the *Journal of Black Studies*, "Afrocentricity in the MNU" (Covin 1990b). This work led to a grant from the American Philosophical Society and a return to Brazil in the summer of 1992. By this time, in addition to studying Portuguese on his own, Covin had been tutored in conversational Portuguese for a year.

He came back to Brazil with a detailed research program: interviewing MNU members, discussions with Brazilian scholars, access to MNU documents, and extensive participant-observer opportunities. The possibility of realizing this agenda was made feasible only by a continuing correspondence with de Lima, from 1986 to 1992. Two hours after his arrival in Salvador, Covin was on the phone with de Lima, setting up a detailed schedule for the following day to enable the completion of the fieldwork within 4–6 weeks. Without de Lima's support, both before and after Covin's arrival, the completion of this research would have been impossible. During this visit, Covin met, interviewed, and engaged in extensive interaction with Luiza Helena Bairros, the first and newly elected National Coordinator of the MNU, who was residing in Salvador.

During the same stay Covin met Michael Hanchard and Angela Gilliam for the first time. They were part of a Fulbright program spending time in four Brazilian cities. Hanchard invited him to sit in on some of the group's workshops and seminars. Kim Butler was also in Salvador at the time doing fieldwork for her dissertation. Covin saw her at a *Candomblé terreiro*, but did not meet her until nine years later. *Candomblé* is too complex a subject to explore here, but it serves to note that it is a religion of African origin, which, particularly in Salvador, is associated with West African Yoruba beliefs and practices, even though there are many expressions of *Candomblé* in Brazil.

Hanchard was on the faculty of the University of Texas, Austin. He hosted a conference entitled "Racial Politics in Contemporary Brazil." He later edited a book of the same title, based on the papers presented at the conference (Hanchard 1999). While Covin was unable to attend, it inspired him to invite black political scientists interested in Afro-Brazilian politics to present papers on an NCOBPS panel that year. Hanchard did not attend the NCOBPS meeting, but Michael Mitchell did. Morrison attended, but did not present a paper. Ollie Johnson, then a PhD candidate, attended, as did Dianne Pinderhughes. She had attended a meeting of the International Political Science Association in Rio in 1984 and had traveled to Salvador on that occasion. Since then she had had an abiding interest in Brazil. Stimulated by the panel, over the course of the conference, a group of scholars pledged to offer a panel on Afro-Brazilian politics at NCOBPS every year. That NCOBPS meeting, held in Oakland, in March 1993, marked the birth of the Brazil Group. Its core members were Michael Mitchell, Dianne Pinderhughes, K. C. Morrison, James Steele, Ollie Johnson, and David Covin.

The Seed Ground for the Brazil Group

Let us reflect on some of the circumstances, largely serendipitous, which enabled the foundation of the Brazil Group. Three major conceptions from social movement theory readily lend themselves to exploring this terrain: social space, social narrative, and social memory. Social spaces are places (not only physical places, but also conceptual spaces such as books, movies, and cyberspace) where people can engage questions which concern them and where they can encounter other people similarly disposed. Social narrative consists of messages exchanged in social spaces, and social memories consist of collective memories of important social, political, or cultural events, places, and persons (Evans and Boyte 1985; Couto 1991, 1993; Covin 1997a, 1997b, 2001, 2006, 2009).

With respect to social space, two specific black social organizations (both academic)—NCBS and NCOBPS—provided Morrison and Covin the opportunity to meet and share a similar social narrative rooted in common social memories. It is worth noting that the social memories and narratives they shared had nothing to do with Brazil. The papers Morrison presented were based on his book *Black Political Mobilization, Leadership, Power, and Mass Behavior* (1987). Covin's papers were based on research he was doing on black politics in Sacramento, heavily influenced by Rod Bush's book *The New Black Vote* (1984). Both scholars were steeped in the social narratives arising from social movement theory. Each was also, unbeknownst to the other, curious about Brazil and eager to learn more about it, and thus each had independently proposed to submit a paper to the Association of Caribbean Studies Conference in Bahia. While the Association of Caribbean Studies was not an explicitly black organization or social space, its subject matter was overwhelmingly black, and the dominant social narratives in the organization centered on African descendants. Brazil's huge black population had stimulated both men's curiosity about the country.

In Bahia, Afro-Brazilians were drawn to the Association of Caribbean Studies Conference by the extraordinary number of black scholars who were attending, most from other parts of the Americas, and particularly because they were coming to conduct serious discussions of African-descended peoples. In 1986, Brazil was still in a period labeled the *abertura democratica*, a stage marking the period of liberalization, a *perestroyka*, if you will, when arrangements were being implemented for the transfer of political power from military to civilian authorities. Indeed, an electoral college whose membership was largely determined by the military had elected a civilian president, Tancredo Neves, who died almost immediately. He had been succeeded in office by his vice president, José Sarney, a favorite of the military, though also a civilian (Mitchell). The government, nevertheless, was still overwhelmingly dominated by the military, that is, while nominally under a transitional government, the country still operated as a military dictatorship. Here, in this period of new openings, and expressly new social movements, the Association of Caribbean Studies Conference was a space to attend, participate in, and practice this new phenomenon of democratic activity. If one were black, one could even for a while savor the experience of relaxing in a luxury hotel lounge, sitting at the bar, and eating in the restaurant. If heaven had not yet arrived, it nonetheless could not be too far away. Afro-Brazilians were open, eager, and able to engage their counterparts from distant lands.

A word is now in order about just who these black Brazilians who showed up at this event were. Let us start with the two young men who were so bemused because they had a window to get into this first-class hotel and enjoy its amenities.

De Lima was an unemployed civil engineer. He had worked at his profession for several years, but was then out of work. He had a brother who was also a civil engineer, but who had solved the unemployment problem by moving with his family to Angola, where he was successfully employed in his profession. De Lima's other brother was a dentist in Salvador. Their father had died. De Lima lived with his mother in a house she owned in the mid-scale neighborhood of Fazenda de Garcia. Da Silva was a college graduate. He had recently been hired by the national government and was preparing to leave for Brasília where he would start his new job.

While Morgan was white, her husband was a black university professor from the United States, and her children were black. She was a *Candomblé* initiate and identified with the black movement. She knew she was not black and did not pretend to be, but she fully understood the struggle and supported it.

Bairros was not only the National Coordinator of the MNU, she was also a university graduate, had a master's degree, and worked both as a social worker and a university professor at the Catholic University of Salvador. Her husband, Jonatas Conceição, was a member of the MNU, college-educated, a journalist, and a popular DJ. Jonatas had a sister, Ana Célia, who had met Covin at the Association of Caribbean Studies Conference. She was a college professor, a writer, and an activist who also belonged to the MNU. Many who attended the conference had similar backgrounds. There were many *Candomblé* adepts and important community figures. There were cultural leaders and artists, teachers, university professors, and college students. Movement organizations were well represented.

These were not residents of the *favelas*. Yet, without the presence of the conference, they would not have been allowed inside the hosting hotel if it were not for the conference—even though many of the hotel's employees were black—and not only housekeepers, janitors, cooks, and pool attendants, but also people at the registration desk, and office workers.

The people who met and befriended Morrison and Covin were Afro-Brazilian elites. By the country's standards at the time, they were well-off, particularly for black people. They were also members of organizations. They had constituencies. They were not isolated. They were opinion leaders. They tended to have connections with their white colleagues—those who controlled money and power. Whether through professional relationships, common *Candomblé terreiro* memberships, friendships, or even family relationships, they were able to leverage such linkages for their own priorities, which were usually centered on race, the movement, and organizations. The US scholars' contacts, acquaintances, and friends enabled them to operate in a privileged context, although they did not understand that.

The Brazil Group

In the summer of 1993, Bairros came to the United States to study in the African Diaspora Program at Michigan State University (MSU) and earn a PhD in sociology. She had maintained contact with Covin through mail and telephone. During the fall semester Covin invited her to come to NCOBPS, to attend the Brazil Group panel, and to participate in the group. She accepted the invitation. Her participation marked changes in the Brazil Group that would eventually lead to the Race and Democracy Project. NCOBPS met in Hampton that year—one of a number of events that were held at Hampton University—and that had a significant impact on Bairros.

After the first full day of conference events, buses took NCOBPS participants to the Hampton University campus for dinner. The buses stopped at the campus entry where Bairros noticed a large plaque that commemorated the institution's founding in April 1868. She was dumfounded. She said: "In Brazil, slavery still hadn't ended. It would go on for twenty more years. Here there was already a black university. No wonder we have so much work to do."

The Brazil Group met for long hours and frequently at that conference. On a Saturday evening, after the meeting's official close, on a second-floor balcony of a restaurant in Hampton, they spun pipe dreams about all coming together at a conference on the beaches of Salvador. At subsequent NCOBPS conferences the Brazil Group continued to sponsor panels on Afro-Brazilian politics. It expanded the concept to include panels on the Caribbean and Afro-Latin America. In settings away from the conference panels, largely hotel lounges and restaurants, its members maintained the practice of holding separate meetings, discussions, and entertaining ideas for collaborative work. One idea that surfaced was an edited book, with chapters written by various members of the group. The group also sponsored panels at the American Political Science Association and the Midwest Political Science Association.

In all these deliberations, Bairros articulated the lamentable state of black graduate students in Brazil. Few Afro-Brazilians attended college. Once there, they lacked the resources to stay enrolled on a continuous basis. They pieced their degrees together over decades, working part-time, dropping out to work full-time, and then going back to study. Research funds were completely controlled by a small group of white scholars. Brazil's equivalent of the National Science Foundation (NSF) only accepted proposals from PhDs. There were so few black PhDs, especially in the same field, that they were unable to mount projects that had a chance for funding. These considerations increasingly influenced the group's discussions. Gradually conversations turned toward examining what the group would propose if it were to seek funding.

Bairros was insistent that whatever they did should be collaborative, such that Brazilians and NCOBPS members would be coequals. It would have to be interdisciplinary, as she knew of no Afro-Brazilians who had completed the PhD in political science. It would have to be funded from the United States because there were no Afro-Brazilian scholars who had direct access to funding sources. Further elements were discussed including language training; the need to incorporate both junior and senior scholars; the formation of mentor relationships—cross-nationally as well as intra-nationally; the development of cross-national research projects; the development of support mechanisms for junior Afro-Brazilian scholars; and the implementation of mechanisms to insure that both the conception and inception of the work would be cross-national. The vision and design would derive from cross-national collaboration, deliberately including scholar-activists as full-fledged members of the project.

The Brazil Group decided to submit a grant proposal to the NSF through NCOBPS. Work began on drafts. In 1997, Bairros finished all her course work at MSU and passed her comprehensive exams. Her dissertation proposal was approved, and she returned to Brazil to complete her research and writing. For whatever funding proposals might emerge, she was designated as the Principal Investigator from Brazil. All of this work was enabled by the social space afforded by NCOBPS, and the social memory and narratives afoot there.

Morrison and Covin were designated as co-PIs in the United States. A planning grant proposal was submitted to the NSF, through the University of Missouri, Columbia, Morrison's home campus. In the fall of 1998, that grant was funded. In January 1999, Morrison and Covin went to Salvador to begin planning with their Brazilian cohorts. One of Bairros's many roles in Brazil was as a researcher for the Center for Human Resources (CRH) of the Faculty of Philosophy and Human Sciences (FFCH) at the Federal University of Bahia (UFBA). The Center made its facilities available for the planning meetings, which lasted four days. Two plans emerged. One was a bare-bones proposal designed for the NFS. The other was the dream plan, incorporating a full realization of the planners' aspirations.

The bare-bones proposal was submitted to the NSF, which denied funding, but made recommendations for resubmission. Meanwhile, Dianne Pinderhughes had been holding talks with the Ford Foundation about the proposal, particularly with Margaret Wilkerson, the then Ford Program Officer in Education, Knowledge, and Religion; Education, Media Arts, and Culture. Wilkerson agreed to discuss the proposal at the Stanford Conference on Race: African Americans: Research and Policy Perspectives at the Turn of the Century, to be held in November 1999 (hosted by Lucius Barker, a former NCOBPS president). At the conference, Wilkerson expressed interest in the proposal, and invited the Group to submit it under the auspices of NCOBPS. When asked whether to submit the bare-bones or the dream proposal, she said the Foundation would be interested in seeing the Group's fullest conception. It was submitted through the California State University (CSU) Sacramento Foundation in December 1999. While the proposal was under consideration, Wilkerson made a site visit to CSU Sacramento. There, she had a meeting, attended by the co-PI (Covin), with the university president, Don Gerth; the Dean of the College of Social Sciences and Interdisciplinary Studies; the Director of the Office of Research and Sponsored Projects; and the Office of Global Education. All expressed enthusiastic support for the project. The Ford Foundation announced funding of the entire project in March 2000. NCOBPS operated as a subcontractor.

The Race and Democracy in the Americas Project

To meet some of its long-term objectives, the Brazil Group had recruited four junior scholars. Applications were solicited from NCOBPS graduate students and recent PhDs. Extensive formal evaluations of the applications resulted in the selection of Hawley-Fogg Davis, a new Princeton PhD and assistant professor at the University of Wisconsin, Madison; Mark Sawyer, a new University of Chicago PhD and assistant professor at UCLA; Heather Dash, a graduate student at Emory; and Rosalind Fielder, ABD from the University of Illinois, Urbana Champaign, who had just received a dissertation fellowship from Vermont University.

Meanwhile, Bairros was busy in Brazil. She recruited twenty-five Brazilians, one Uruguayan, and one Cuban, by way of Jamaica. Six of her recruits were PhDs. The rest, including Bairros herself, were junior scholars. The Brazilians came from many locations, including Salvador, Brasília, São Paulo, Recife, Rio de Janeiro, Belo Horizonte, and Feira de Santana.

The First Convening of the Project

All the participants first came together during May 22–26, 2000, at a *Seminário* in Salvador and Itapuã (a Salvador suburb). It was an event of extraordinary significance. It was the first all-black *Seminário* of its kind in Brazil among Afro-Brazilian scholars and activists and between US black political scientists and Afro-Brazilian social scientists. There had been many academic conferences in Brazil on black culture, the black movement, black history, race, racial politics, and related subjects in the past, but the over-whelming number of participants was white. At fora such as these, which should have seen significant black participation, there were at the most one or two Afro-Brazilians on a few panels. The black voice was barely present. Thus, the 2000 *Seminário* was unprecedented. Indeed, it represented a rare opportunity for Afro-Brazilian scholars to come together from all over the country to discuss common concerns and interests. It would not be an exag-geration to say that they were more excited and enthusiastic about meeting with each other than with their North American counterparts. This was a new kind of space in Brazil where Afro-Brazilian scholars were free to con-duct social narratives using their own social memories, without outside influ-ences. The *Seminário* took place in two parts spread over five days. Part one was a one-day public conference at the Hotel da Bahia Tropical. The opening speakers included representatives from the CRH, FFCH, UFBA, the State University of Bahia (UNEB), the Center for Afro-Oriental Studies (CEAO), NCOBPS (Morrison was then the president), and Wilkerson, representing the Ford Foundation.

Also on the program was the former and first black governor of the state of *Espírito Santo*, Albuíno Azeredo; representatives from Belo Horizonte; the Federal Congress; the black movement, including the MNU; and the black press. The event was open to the public and there were spirited discussions and Q&A sessions. It lasted from nine in the morning to six in the evening.

Part two took place in *Itapuã*, north of Salvador, in a former convent con-verted into a Leadership Training Center, with a beach on the Atlantic. This was the working part of the conference for the academics. It started immedi-ately after breakfast and lasted all day with a lunch break in between. For the first two days there were presentations on ongoing or completed research work, with plenty of discussions and Q&A sessions that went on all day, frequently extending into the evenings. The Brazilians stayed up most nights talking until two or three in the morning; few if any of the US participants could keep up with them. Seventeen papers were presented during the first two days. The second two days were devoted to discussing possible collabora-tive studies, and finally to establishing the framework for the second convening of the project which would take place in Sacramento, California, during July 7–13, 2001. This was a new phenomenon—an international black space, based in the western hemisphere. Its participants cross-fertilized their social narratives and reshaped their social memories. It was an extraordinary eye-opening experience.

The whole *Seminário*, but particularly the first day's open public conference, received extensive press coverage from both the electronic and print media. It made first-page news and featured articles in all four of Salvador's major newspapers. It also was featured on both television and radio news locally, with many live interviews. There were several preconference articles about the upcoming event, as well as postconference follow-ups.

The Second Convening of the Project

The second conference and workshops followed the same scenario as the first. There was an all-day conference open to the public on the first day, followed by four days of intensive workshops. New participants from the United States were invited for these sessions. Among them were Kim Butler from Rutgers; Anani Dzidzienyo from Brown; Rovana Popoff, a graduate student from the University of Chicago; Micol Seigel, a graduate student from NYU; Denise Ferreira da Silva, an assistant professor from UC San Diego; and Raquel Souza, a graduate student from the University of Maryland, College Park. Guest speakers included, among other local notables, the Brazilian Consul General from San Francisco; José Augusto Lindgren Alves; Maulana Karenga; from US and CSU Long Beach; and the future Mayor of Sacramento, Kevin Johnson.

The significance of the Sacramento Conference was perhaps put most clearly by Anani Dzidzienyo, a pioneering scholar and mentor for black scholars studying Afro-Brazilians, whether from the United States, Brazil, or elsewhere. He said that at the path-breaking conference at UCLA in 1980 which produced *Race, Class, and Power in Brazil* (Fontaine 1985), one Afro-Brazilian scholar had been present (Lélia Gonzalez). Since then, there had been two or three other conferences with a similar focus. For each, two to three Afro-Brazilian scholars traveled from Brazil. He said that in 1984 he could not have imagined that more than 20 Afro-Brazilian scholars, most of them coming directly from Brazil, would attend a single conference in the United States for a scholarly examination of the racial question. Even in 1991, it would have been unthinkable. The unprecedented Sacramento Conference was both shocking and inspirational. It was something he had not expected to see in his lifetime. The international black space, realized in Salvador, had set itself up in Sacramento.

The opening day of the conference received nothing like the scope and thoroughness of press coverage present in Salvador. The exception was the local black newspaper *Sacramento Observer*, another black space, which featured a seven-page supplement, which not only covered the conference in depth, but also used it as an opportune teaching moment about Afro-Brazilians, including demographics of Brazil, black elected officials in Brazil, and a feature on Salvador and Rio de Janeiro. This was a different kind of social narrative from that found in the popular press. The four days of intense workshops proved as effective as those in Salvador. Again, there were 17 papers presented at the workshops. Reports were made on collaborations under way, and more time was spent on brainstorming about planning further collaborative efforts.

During the second and third years of funding, both US and Brazilian participants were frequently brought from one continent to the other for conferences, probably more to NCOBPS than to any other, for workshops, and as visiting lecturers or scholars, perpetuating the international spaces and narratives, creating new kinds of social memories (Table 2.1).

The participants in such intense and longitudinal relationships could not help developing new kinds of social memories.

Table 2.1 Achievements through June 2002

Number	Category
1	MA theses under way, accepted into PhD Program (United States), PhD completed (Brazil)
2	Supported major international conferences; articles submitted for publication, still under consideration; journal editions based on project work (one in the United States, one in Brazil); project members awarded tenure at major research universities (United States)
3	Project-related books published, MA degrees completed (one in Brazil, one in the United States)
5	Initiatives developed at universities, ongoing programs strengthened
6	Participants engaged in language study
7	Proposals funded, research projects developed by individuals
8	PhD dissertations begun (five in Brazil, three in the United States)
9	Funding proposals initiated by individuals
13	Invited lectures, mentor relationships
32	Project-related articles
51	Appointments and elections to official positions
52	Project-related presentations at scholarly conferences
60	Project-related appearances at non-scholarly events

Note: This listing identifies the project's quantifiable accomplishments by June 2002.

Subsequent Funding

In 2003, the funding for the Race and Democracy Project was almost doubled, and was extended through the end of 2005. (The asterisks "*" indicate the involvement of black social spaces.)

Major Initiatives

Two recruits from Ecuador recruited to the project and brought to NCOBPS meetings.*

English and Portuguese language workshops in Salvador every summer, staffed by Raquel Souza, a project participant.*

Brazilian junior scholars enrolled in Language Institute classes in English during the school year in Salvador, São Paulo, and Brasília.

Summer methodology workshops conducted by Mark Sawyer and Dianne Pinderhughes.*

Work with the Steve Biko Cultural Institute to train students for the *vestibular*.*

NCOBPS members initiated, attended, and conducted studies of the first two meetings of African-descended legislators in the Americas, also worked with the Congressional Black Caucus Foundation on this project. The meetings took place in Brasília and Bogota. Major Project scholars participating were Luiz Alberto, Luiz Claudio, Cloves Oliveira, James Steele, Ollie Johnson, Michael Mitchell, and K. C. Morrison.*

Continual opportunities for scholars based in Brazil to present and give lectures as visiting scholars on campuses in the United States.

US scholars were invited to present and give lectures at both nonacademic and academic settings in Brazil.

Three PhDs completed in Brazil: Economics (*summa cum laude*), Political Science, and Public Health.

One MA completed in the US major conference hosted at UCLA, "Racial Democracy in the Americas."*

Student accepted to PhD program, U. T. Austin. Permanent sites established for the Race and Democracy Project: UFBA, Brazil; UCLA, United States.*

And Beyond

The last external funding for the project ended in 2005, yet the project continues. Since 1992, NCOBPS has never failed to offer at least one panel on Afro-Brazilian, Caribbean, or Afro-Latin politics. Junior NCOBPS scholars continue to be mentored and encouraged in their work on the politics of the

African diaspora. Among such scholars are Savannah Carol, Jaira J. Harrington, and Tonya Williams (now a PhD on the faculty of Johnson C. Smith University).

On the Brazilian side, the results are even more dramatic. When the project began, 17 participants were either graduate students or undergraduates. All now have an MA. Some have multiple MAs. A minimum of seven have a PhD. Five are ABDs. Ironically, the most critical founding members on the Brazilian side, Bairros and Luíz Cláudio Barcelos, did not complete their PhDs, though they are ABDs. Bairros was appointed by President Dilma as the National Cabinet Minister in Charge of the Special Secretariat for the Promotion of Racial Equality (SEPPIR), a position she held for four years. Luiz was her Deputy. Seven Brazilian project members have published books. Several have published more than one. Of the seven who have published books, three were graduate students when the program began, and one was an undergraduate. The Brazilian participants, collectively, have written scores of articles and book chapters. One project member, Luíz Alberto, has been elected to the National Congress for multiple terms. Silvio Humberto Cunha was recently elected to the Salvador City Council. For the Race and Democracy Project, it is noteworthy that upon Bairros's appointment to the Cabinet, the National Health Council, a black health advocacy organization, began its citation of her achievements thus: "She participated in coordinating the research of the Race and Democracy in the Americas Project: Brazil and the United States, a project concerned with racial discrimination" (*Saúde Negra* 2011).

Similarly, the Voice of Black Brazilian Women began their citations noting that "she founded, in partnership with the National Conference of Black Political Scientists (a North American Organization), the Race & Democracy in the Americas Project, that promotes exchanges between Afro-Brazilian graduate students and researchers and Afro-North Americans" (*Família* 2012).

The achievements noted above, along with the other work and accomplishments of project members, as well as the work of today's new graduate students who have no knowledge of the project at all, are built—and will continue to build—on the foundations created by the project. These achievements are roads paved by the footsteps of—but that go beyond—the Race and Democracy in the Americas Project, roads erected in black spaces, creating new kinds of social spaces, new kinds of social narratives, new kinds of social memories, and new kinds of scholarship.

CHAPTER 3

Brokering Black Brazil or Fostering Global Citizenship? Global Engagement that Empowers Black Brazilian Communities

Elizabeth Hordge-Freeman

Introduction

I waited somewhat impatiently in a perpetually long line to use the restroom at Sankofa African Bar & Restaurant in Salvador, Bahia, Brazil, and when I glanced around, I saw two black women behind me whispering to each other and smiling. I later learned that one of the women was Brazilian and the other was African. Quite casually, one of them asked me (in Portuguese): "Where are you from? Are you from Brazil?" Before I could answer, the other smiled and chimed in: "See, I think she's African, maybe from . . . Angola?" The other responded quickly: "Look, she has a long face, and that forehead." The other interrupted: "But her lips (*looked at my lips and frowned*) are too . . . (*shook her head no*)." Curious about the interaction, I remained quiet and smiled coyly, refusing to speak in order not to reveal my accent and/or my identity. When the anticipation (and my discomfort with them deconstructing my facial features) was too much to bear, I revealed: "I'm from the United States." To which one of the women said to the other: "See I told you! But, you (*referring to me*) look like you could be from here [Brazil]." This brief interaction alludes to what happens when diasporic groups meet: misidentifications may complicate coalition building, but expectations of solidarity and similarity are also suggestive of the possibilities of diasporic engagement.

This chapter explores the intricacy of diasporic engagement, by highlighting the opportunities and challenges involved in developing the University of

South Florida (USF) in Brazil summer program.[1] To organize the chapter, I use a critical feminist perspective and asset-based approach, which provide frameworks for assessing the program. I discuss ways in which my positionality as a black researcher from the United States provided me access to important research networks and key persons in Brazil who could facilitate the development of the program. But, my inclusion in these transnational networks was not automatic—it was the result of friendships and my participation in reciprocal research and teaching partnerships. Moreover, my inclusion required (and continues to require) that I take seriously power differentials and the privileges that come with my Americanness in order to forge truly collaborative projects. Rather than having a discrete end point, this process must be continual and critical. The chapter ends with conclusions for scholars who hope to further develop their research and teaching in ways that will reflect a commitment to diasporic engagement and (critical) global citizenship.

The city of Salvador is an inevitable stop for researchers who are interested in exploring blackness in the African diaspora. A homage to its significance, at various points in history, Salvador has been considered Black Rome, an idealized destination for black folks (especially those from the United States) seeking to reconnect with their African roots (Pinho 2010; Ickes 2013). The relevance of Salvador, in particular, in the diaspora is also embodied by how blacks from across the diaspora meet and interact in social, political, and intellectual spaces around Salvador. As a black researcher from the United States who conducted research in Salvador, from 2009 to 2010 and in the summers from 2013 to 2015, one of the spaces that emerged from these diasporic connections is Sankofa African Bar & Restaurant.[2] Sankofa is intriguing because it was there that I witnessed the multidimensionality of diasporic interactions unfold through discursive exchanges such as the one mentioned in the introduction.

As a constructed diasporic space, Sankofa shifted from being a delicious eatery to a space of political engagement to a place that served as an intellectual haven for racial consciousness-building. I attended a spoken word jam session there where *negros assumidos* (Brazilians who assume their blackness) roused an attentive crowd by using prose to crush racial constructions and challenge racial exclusion. Besides these political encounters, Sankofa hosted diasporic cultural exchanges, and it was here that I (as well as other blacks from Brazil, several countries in Africa, and those from the United States) spent nights dancing merengue (traditionally from the Spanish-speaking Caribbean), practicing samba (a traditionally black Brazilian dance form), and listening to music from the United States and Senegal. Drawing on what Sankofa had seemingly perfected, I envisioned developing a USF in Brazil summer abroad program that would similarly use the commingling of entertainment and diasporic engagement to promote intellectualism, cultural criticism, and political involvement.

Limitations and Possibilities of Global Citizenship

There is an extensive historical trajectory of black scholars conducting research in Brazil and the emergence of a new wave of interest in fostering "global citizenship" and "global engagement" provides an opportunity for diasporic researchers to bring their interest in Brazil to a broader audience (Hellwig 1992; Matory 2006). But this notion of global citizenship does not begin or end with college students or Brazil. The growth of global engagement in universities is preceded and accompanied by the popularity of "voluntourism," which refers to short-term programs in which "tourists . . . volunteer in an organized way" and "undertake holidays that might involve aiding or alleviating the material poverty of some groups in society, the restoration of certain environments or research into aspects of society or environment" (Wearing 2001, 1). While on the surface it may appear to be harmless, voluntourism has been scathingly critiqued on the basis that it targets vulnerable communities and can be paternalistic in ways that "exacerbate inequalities and cultivate dependency" (McLennan 2014, 165; see also Devereux 2008; Scheyvens 2011). In some cases, these voluntourism programs can actually have detrimental effects on the local communities that they are meant to help (Epprecht 2004). Providing an analysis of the egocentric motivations of volunteer tourists, Sin (2009) finds that "many volunteer tourists are typically more interested in fulfilling objectives relating to the 'self'" than actually improving the conditions in which they are working (497).

In this sense, the truly transformative potential of global citizenship and community engagement can be co-opted by programs that promote new forms of imperialistic relations that dismiss the actual needs of local communities (Caton and Santos 2009; Doerr 2013; Hartman et al. 2014). Practically, some researchers argue that volunteers in short-term programs are often not able to have a meaningful impact on social problems in the areas that they target, but "by parlaying their economic capital into symbolic capital through the purchase of a volunteer vacation, already privileged people are able to appear distinctly altruistic and worldly" (Heath 2007, 237; see also Simpson 2004). Ironically, after handsomely investing in the "feeling" of having done their part, many privileged do-gooders return to a life of insularity where ideas of altruism are not extended to poor people (let alone poor people of color) in their native countries.

Despite the numerous ways that global engagement and global citizenship have been appropriated, I hoped to use my positionality and expertise to develop a program in Brazil that would not fall prey to these same pitfalls. Indeed, the increasing development of programs that promote global citizenship at universities, including the study abroad and global service-learning programs that are becoming staples of the college experience, suggests that universities have adopted the same optimism. In fact, study abroad and global engagement are so ubiquitous that private universities where endowments can fund these endeavors "all but require" these global activities and have begun to reorganize institutional structures to accommodate these changes (Looser 2012).[3] College courses designed to incorporate global themes also point to the way that global citizenship is being integrated into traditional pedagogy. My critiques of global citizenship, rather than being a reflection of my fundamental problem with the notion, emerge because I am the product of the institutionalized investment in global engagement and research in Brazil. My aim with developing the USF in Brazil program has been to draw on my positionality as a black researcher from the United States to build the type of community partnerships that can promote ethical and critical global citizenship in the diasporic context of Salvador.

Conceptualizing Global Citizenship in Salvador

Researchers would benefit from more transparency regarding the development, implementation, and evaluation of international summer programs (see Bell and Anscombe 2013; Solís et al. 2015). With this goal in mind, I examine the achievements and limitations of the USF in Brazil global service-learning program, which I launched and directed in 2013 and 2014. I relied heavily on the principles of critical feminist perspectives to develop the program. The critical feminist tradition places emphasis on cultivating specific practices, discourses, and relationships in order to promote both engagement *and* empowerment of vulnerable communities (McClaurin 2001; Collins 2000). These philosophical commitments require mobilizing diasporic connections in Brazil in ways that do not allow unequal power dynamics to be hidden under superficial engagement that is disguised as "altruistic" (Jakubiak 2012, 436). Aligning my work with this critical perspective also means that I recognize that I must continually work to reject, on the ideological and practical level, exchanges and interactions that might make me complicit in callously brokering black Brazil. The term "brokering" is particularly relevant to the Brazil program because it captures how the program's academic offerings activities were inevitably tied to the cultural commodification of Afro-Brazilian culture (Pinho 2010). This does not mean that I was necessarily complicit, but rather that the program was developed in a context where certain elements of black culture were packaged for an international audience.

Even though in the description of the curriculum, I explained that the program would problematize the commodification of Afro-Brazilian culture in Salvador, it was still engaged with a particular marketing of Salvador as a city that should be visited and studied.

My identification as a black woman did not automatically make me sensitive to questions of power and equity. Even when I did have heightened sensitivity to certain issues, this did not mean that I knew how to best handle or correct situations. I, too, needed guidance about how to plan activities for students that showed the complexity of black life in Salvador and that allowed for a portrayal of black Brazilians as multidimensional citizens, rather than simply as entertainers or mere victims. This is no small task in a city where there is a booming tourism industry that is based largely on the commodification of blackness. Further complicating this is the fact that many black Brazilians who are left with few options make a living by performing stereotypical ideas about blackness in Bahia (Pinho 2010). By consulting with local community groups and following models developed by other programs, the Brazil program would draw on local knowledge in order to resist cultural commodification and construct mutually beneficial experiences for Brazilian and US students that were consistent with critical global citizenship.

With my research and teaching approach, critical global citizenship can be evaluated, in part, based on whether research and teaching includes multiple voices, addresses community-defined problems, and recognizes the agency of community members. The Brazil program emphasizes race and gender, and relies on a critical feminist pedagogical approach to facilitate students' understanding and engagement with questions of racial and gender inequality (Luke and Gore 1992). This cultivates a safe intellectual space in which students are expected to contribute their thoughts and respectfully discuss the perspectives of others. Another major element of a feminist pedagogy is that it relies heavily on active learning through participation in local events and global engagement, such as global service-learning programs (Chow et al. 2003). Destabilizing traditional professor/student dynamics, class time and discussions are geared toward having students reflect on their experiences while discussing strategies to combat racial and gender inequality (Chow et al. 2003; Costa and Leong. 2012).

In developing the curriculum, I relied on an "asset-based" approach, as the course relates to working with global partners. This approach illustrates that by building upon the knowledge and skills that are present in the community, researchers gain valuable insight into how programs can best achieve community-defined objectives (Kretzmann and McKnight 1993; Mancini et al. 2005). One of the outcomes of collaborative interactions is that researchers engage community groups as experts and partners, rather than as merely subjects.

Because mutually beneficial outcomes are the goal, constant self-reflexivity and ongoing discussions with community partners are critical to the process and help to ensure reciprocity. Indeed, reciprocity does not have an end point, but instead it is constantly managed and renegotiated as a reflection of the dynamic nature of community relationships and capabilities (Dostilio et al. 2012). The importance of reciprocity in global community engagement takes on even greater urgency because there is often little researcher ability or oversight in international settings, especially when it involves Western researchers (Mohanty 2003).

Though I had hoped that research reciprocity would organically occur, what I consider reciprocity is constantly being reevaluated (Hordge-Freeman 2015a). My efforts to demonstrate reciprocity emerged as early as my dissertation research in Salvador from 2009 to 2010. As a doctoral student I served as a volunteer English teacher, wrote college recommendation letters for students, and provided free GRE tutoring to Afro-Brazilian students from Instituto Cultural Steve Biko and the Federal University of Bahia (UFBA). This was the very least I could do considering all that Afro-Brazilian scholars and community members had done to assist me with my research. Now, as an assistant professor, my research agenda has developed considerably and so has my conceptualization of reciprocity, which now involves much more than simple exchanges. Dostilio et al. (2012) elaborate on a multileveled conceptualization of reciprocity called "generative reciprocity," which describes how through "collaborative relationships participants (who have or develop identities as co-creators) become and/or produce something new together that would not otherwise exist" (110–111). This particular type of reciprocity recognizes the knowledge and agency that local community members have to articulate their needs *and* the importance of intentionally cultivating opportunities for their voices to be heard beyond the scope of my formal research. Reciprocity destabilizes hierarchies of knowledge and troubles the distinctions between researcher and subject by promoting shared authority in framing the research problem, collaborating on the approaches to community-defined needs, and even inclusion in data collection (Frisch 1990).

In the light of the aforementioned perspectives, we tried to develop the USF in Brazil program with the goal of integrating what I refer to as the four Cs of critical global citizenship: community-centered, collaborative, critical, and continuous. Community-centered refers to the importance of community partners in determining what type of global service-learning projects should be developed, in order to ensure that they respond to real community needs. Collaborative refers to the way in which service-learning participants and directors should engage with community partners as local experts. Rather than assume that foreigners will lead, there needs to be a willingness to listen to better understand how and why one's research and teaching can or cannot contribute to communities' goals. Criticality refers to the importance of acknowledging the significant privileges that come along with one's positionality *and* being willing to use knowledge gained in global contexts to shape one's advocacy for groups in both global and local contexts. Finally, continuity refers to the importance of working to foster sustainable relationships and/or sustainable outcomes.

Easier Said than Done? Fostering (Critical) Global Citizenship in Salvador

Critical feminist pedagogy paired with an asset-based approach and a strong sense of reciprocity might theoretically be considered the perfect formula for conducting "good" research and developing a "good" study abroad program. But ensuring that these concepts were thoroughly integrated into the USF in Brazil program was a difficult process. If partnerships and relationships are key, how does one actually begin to foster relationships that can pave the way to study abroad programs? In what ways might my unique positionality as a black female researcher in Salvador (considered the cultural center of the African diaspora) either hinder or help these partnerships? How do I know which partnerships to develop?

These were the questions I asked as a PhD student planning to conduct research in Salvador. In 2009, when I was just beginning my fieldwork, a serendipitous invitation to meet the US Ambassador to Brazil in Salvador led me to attend an outdoor dinner event. At the dinner, a white Brazilian man approached me to learn about what I was doing in Brazil and the conversation quickly devolved into his confident assertions that racism does not exist there. As he left my table unfazed by my spirited rebuttals, another made his way over, and in sheer exasperation with my previous conversation, I dismissed him by association. I dispassionately answered his questions about my work, and only became more interested when he conveyed his own frustrations with racism in Salvador. As a Spanish expatriate, he reflected on the ravages of racism in Brazil and shared his vision of working with community partners to create programs to train black Brazilian youth. I found his sincerity and passion compelling enough to convince me that I wanted to learn more. Perhaps an unlikely ally, the Spaniard, Javier Escudero (founder of Brazil Cultural and former Penn State Associate Professor of Spanish) and I have been colleagues ever since. He and I have worked together with local community organizations on numerous projects including the development of the USF in Brazil program. But, while Javier and I laugh each time we tell this story to others, the encounter is a cautionary tale. It exposes my own blind spots and the relevance of how misidentifications can complicate coalition building among those in the African diaspora and also, as I had learned, potentially undermine coalitions involving blacks and white allies.

Drawing on four years of work with Brazil Cultural and my ongoing relationships with black Brazilian activists, friends, and colleagues who were trained at Instituto Cultural Steve Biko, we developed the USF in Brazil program in 2013. Important to mention is that these meaningful relationships with Brazilian activists, scholars, and researchers did not occur immediately, but rather resulted after several years of interactions (Hordge-Freeman 2015b).[4] The main goal of the Instituto Cultural Steve Biko and its affiliate, Quilombo Ilha, is to offer young Afro-Brazilian students a preparatory course to prepare them to pass the university exam. Given their mission, the prospect of developing a partnership with them was appealing. Brazil Cultural had already developed some service learning initiatives with these local organizations, organized seminars, and provided internships to their students. In 2014, we collaborated to convert the USF in Brazil program into a global service-learning course involving Instituto Cultural Steve Biko, Brazil Cultural (a formal USF World partner), and Quilombo Ilha.[5] Furthermore, Jucy Silva, Executive Director of the Instituto Cultural Steve Biko, submitted a letter to endorse a grant proposal, which I was ultimately awarded from USF's Office of Community Engagement & Partnerships, to fund certain elements of the service-learning program. By using existing strengths and knowledge of these organizations in Salvador, I received valuable insight into how to frame the service-learning component of the program so that it could best achieve one of the Institute's major objectives: English language acquisition.

Yet, the implementation of this program was not without challenges. Some colleagues scoffed at the time that I spent organizing and recruiting for the program, referring to the program's development as "service" to the university and, hence, discounting its teaching and research value. A few, who were well-meaning, joked that the Brazil program was a vacation or clever ploy designed to allow me to sit on the beach sipping *caipirinhas*. Moreover, I was often warned that assistant professors (and certainly black assistant professors) should not launch time-consuming study abroad programs. So ubiquitous were these latter comments that I did not tell my closest advisers about the program until after its launch. And, when I did disclose it to them, I was certain to discuss it as supplementary to the most valued academic currency: research publications.

As an "outsider-within" in the academy, I had heard too many horror stories about faculty of color and women who had been denied tenure to simply ignore the reservations of my mentors (Collins 1986, S25; Gutiérrezy Muhs et al. 2012). Yet, I was absolutely convinced that I had the energy, time, and obligation to take risks, especially those risks that would help advance the significant work already being done by black community organizations that had supported my research (Shayne 2014; Williams 2014). As Collins suggests, for women in the academy "marginality" can be the impetus for "creativity," and the development of the USF in Brazil program represents how this idea has materialized in my career (1986, S15). Ultimately, I reconciled my reservations about launching the program by deciding that I would marshal global partnerships and the service-learning program in Bahia, Brazil, in ways that could enhance my research. In this sense, the USF in Brazil service-learning program would not jeopardize my tenure, because "research and political engagement can be mutually enriching," and instead would strengthen my ability to conduct community-engaged research (Hale 2008, 3). This integration of research, global education initiatives, and teaching has had a material (and positive) impact on the quantity and quality of my research, and it has allowed me to exert agency and freedom in carving out a career trajectory more in light with how I view myself as a scholar-activist.

Engaging Community Partners: "Why Do Americans Waste So Much?"

The USF in Brazil program was conceived as a four-week program, different from others in that it was also a homestay program. Students began the program on the island of Itaparica—an island that is a 45-minute ferry ride from the city of Salvador. This served as an introduction to the complexities of the region and allowed them to observe how the geographic distance between the island of Itaparica and Salvador has had devastating economic consequences for island residents. During this first week, students lived in one large island house, an arrangement which was meant to foster camaraderie among students from different US institutions. Cultural activities including Angolan capoeira lessons with two black Brazilian capoeira instructors at the local capoeira school on Itaparica Island, Anatelson das Neves and Bundião, provided a critical perspective necessary for the students to understand the historical roots of the more popular form of regional capoeira, which they would later see performed in Salvador. Once they arrived on the island and entered their homestay, in addition to their formal classes, students visited standard tourist locations, including the Historic Center (Pelourinho), Mercado Modelo, Pierre Verger Cultural Center, and Balé Folclórico (Folkloric Dance), attended a soccer game and had day trips to Cachoeira and Praia do Forte.

These activities were relatively similar in the USF in Brazil program in 2013 and 2014. However, with the service-learning grant in 2014, USF in Brazil students also collaborated with more than ten Afro-Brazilian service-learning language partners from Instituto Cultural Steve Biko and engaged in interactive activities with over 50 Afro-Brazilian students from Quilombo Ilha. In the context of the program, Afro-Brazilian students who had taken courses at Instituto Cultural Steve Biko were selected to serve as service-learning leaders. They were encouraged to present elements of their culture that are often overlooked or reductively stereotyped in popular culture. The students from Instituto Cultural Steve Biko and Quilombo Ilha shared their life experiences with USF in Brazil students, and were also exposed to other areas of Bahia by participating in the day trips and excursions. These excursions were important because many of the Brazilian students had never and would have likely not visited these locations because of financial constraints. Rather than being passive participants, the Afro-Brazilian student leaders explained this fact as evidence of racial exclusion. Moreover, the community organization Quilombo Ilha facilitated a presentation in which I shared my research on Afro-Brazilian families with Brazilian and US students. After this presentation, we had a question-and-answer session about how racism shapes black Brazilian families. Later in the month, students at Quilombo Ilha organized a traditional party of the northeast (*forró*) in which they cooked food that reflected regional Bahian cuisine and invited USF in Brazil students to learn more about their culture. Indeed, watching the US students struggle to speak Portuguese with the Brazilian students who struggled to respond in English while dancing and eating reminded me of the commingling of culture and politics that had initially inspired the development of this program.

During the USF in Brazil program, US students formally enrolled in two classes: Portuguese language and Afro-Brazilian Culture and Society. The language component of the program was a requirement, which was also useful for their interactions with Brazilian students. Understandably, there are concerns rooted in the ways that global service-learning programs that teach English to non-English speakers might be viewed as a form of cultural imperialism (Jakubiak 2012). Indeed, in Brazil, I had witnessed a cultural and linguistic hierarchy in which the English language and US culture were situated firmly at the top. This overvaluation has always made me uncomfortable, in part because my access to certain networks in Brazil has been shaped largely by the privileges that my Americanness offers (Hordge-Freeman 2015a, 2015b). My Americanness came with a certain cachet and status in Brazil—an odd feeling for blacks from the United States who may feel (and sometimes are treated) like second-class citizens in their own country.[6] I reconciled initial concerns about English language acquisition based on the larger goal of the program, which was not unidirectional language learning, but rather bidirectional exchange. Perhaps the more compelling reason that English made sense is that for the past several years, the Brazilian government has offered substantial scholarship and grants for Brazilians to study in STEM fields in the United States. Because English is a prerequisite for many of these scholarships, many Afro-Brazilians are excluded from such opportunities. Recognizing the value that English training provides both as a pathway to educational opportunity and economic mobility, the Instituto Cultural Steve Biko has worked to develop English language skills.

Though English language has been prioritized, we learned from members of Instituto Cultural Steve Biko that some black Brazilian women students suffer from low confidence and self-esteem in the classroom, which impacts their class participation, especially in the presence of men. This is only one example of the multiple ways black women experience oppression. Their reticence is attributed in no small part to Brazil's *machista* society and the pervasive derogatory stereotypes of black women in Brazilian society (Caldwell 2007). To remedy this, Instituto Cultural Steve Biko developed a women-only English course taught by Raquel de Souza (PhD student from The University of Texas), which was designed specifically to promote Afro-Brazilian women's English language acquisition. Building on the momentum and training provided by this course (a reflection of an asset-based approach), in conjunction with Brazil Cultural, the USF in Brazil program in 2013 and 2014 selected two black Brazilian women who took this course to serve as leaders in our program.

The Brazilian language students had previously taken a course at the Institute Cultural Steve Biko entitled "Citizenship and Black Consciousness," which served as the basis on which the Brazilian service-learning leaders created games, sang songs, made presentations, and facilitated conversations about racial inequality in Brazil. A sincere question from one of the Brazilian service-learning leaders—"Why do Americans waste so much?"—challenged USF students to reflect on their privileges and practices. It also provided a space for Brazilian students to articulate more complicated concepts while developing their English skills. In these informal interactions, I facilitated conversations and assisted with some of the comments that were lost in translation. But my role was secondary, so that the interactions could remain as student-centered as possible. In order to further reinforce the bidirectional and student-directed nature of the program, USF in Brazil students also served as critical collaborators. Their role was not only to immerse themselves in the culture and teach black Brazilian students English, but also to gain insights into Brazilian culture and learn Portuguese. This language exchange ensured mutual benefits, particularly as English training offers cultural and educational capital to Afro-Brazilian students, which can translate into economic mobility.

As the Director of the USF in Brazil program, my stay in Brazil during the summers has afforded me numerous opportunities to foster my critical global citizenship in ways that impact other educational groups. In the summer of 2013, I collaborated with Brazil Cultural and served as the Instructor for the Spelman College-in-Brazil program, and taught their course "Afro-Brazilian Culture." The course involved cross-institutional interactive activities and shared class time between Spelman College (an all-women, historically black US college), USF in Brazil students, and black Brazilian students at Instituto Cultural Steve Biko. Likewise, in the second year of the summer program (which was formally a global service-learning course), similar cross-institutional collaborations were organized involving USF in Brazil and Penn State-in-Brazil program students. The latter, who were studying business and Brazilian culture, were largely composed of white students, while the former were a racially diverse group focused on service-learning and social sciences. Although there was initially some hesitation to combine these groups, at the suggestion of Brazil Cultural, both groups attended several excursions together and ultimately the racial composition of the groups provided dynamic opportunities to expand the contours of what students understood about how race, gender, and nationality shape one's experiences in the diasporic regions.

Growing Pains—Evaluating the USF in Brazil Program

Overall, participants from the United States and Brazil expressed that they enjoyed participating in the program.[7] However, simply being equipped with the handy four Cs of global community engagement did not guarantee that the program achieved all of its desired goals. In order to promote transparency, I outline below some of the major limitations of this abroad program, which I hope will be instructive to others conducting research and developing programs in Brazil. The challenges that I encountered in this program emerged from the fundamental reality that global engagement and programs are "never neutral and that our efforts to learn about and engage with others take place within asymmetrical configurations of power" (Rizvi 2009, 267). Perhaps this is best represented by what my initial experience at Sankofa led me to envision—a program for US students to study in Brazil, rather than a program for Brazilian students to study in the United States.

Firstly, the terms of the structure of the program required that, in order to receive the mini-grant to develop the Brazil service-learning course, USF in Brazil students should complete pre- and post-program surveys about their experiences. This is "best practice" because it allows our office to track how our students viewed their experience to determine its perceived usefulness. Equally important as our students' outcomes is the way the local Brazilian students and community perceived the program's impact. I received positive feedback throughout the four weeks of the program from our community partners;

however, a debriefing and formal meeting with student leaders and the Executive Director at the Instituto Cultural Steve Biko would have led to a better assessment of the program. The inclusion of formal evaluation mechanisms would have been one way to develop measures of reciprocity and accountability, which researchers have noted is often absent in these types of programs.

Secondly, USF in Brazil students were sincerely interested in continuing their relationship with the Brazilian students who served as collaborators in the program. While many of the students are now friends on social media, there was not a more institutionalized mechanism to facilitate or encourage this relationship in a sustained manner. When we returned to the United States, I organized Portuguese language tables, which helped to create some continuity for the US students. However, if the goal is a mutual and student-directed initiative, sustainability might have been better supported by the institutionalization of a Brazil club or related student-led club. Part of the programming of this club might involve regular Skype meetings with Brazilian students to promote the continuity of the English language training.

Thirdly, there were a number of reasons that the focus of the service-learning component was language training. As I mentioned with regard to Brazilian students, English language acquisition has implications for social mobility, and educational and cultural capital. Moreover, the acquisition of English language allows black Brazilians to more effectively leverage international networks in order to promote the change necessary to improve their social, political, and economic conditions. Given the importance of language exchange, US students would have benefitted from basic pedagogical training related to teaching English before leaving for Brazil. As one's status as a native speaker does not make one skilled at teaching English, USF in Brazil students would have been better prepared to address this community-defined need had they received some minimal training in this area.

Additionally, in both the summer programs of 2013 and 2014, I struggled to find a balance between working with Brazil Cultural to plan enjoyable cultural activities and also allow room for critical engagement. While USF in Brazil students were exposed to alternative narratives of blackness in Bahia (largely through their interactions with students and invited speakers), I still found myself wary of certain planned cultural activities. Specifically, I have been uncomfortable taking students to attend Candomblé (African-origin religion) ceremonies because of my concern that it may be a form of cultural voyeurism. To address this, before visiting a *terreiro* (a house of Candomblé), students attend lectures by practitioners and local experts of Candomblé (including presentations by noted oral historian Nancy de Souza e Silva) to learn more about the religion, the Orixás, religious symbolism, and its commodification in Salvador.[8] Yet, even with this preparation before students visited these sacred places (even with an invitation) did not allay my concerns. It is possible that my uneasiness was more a reflection of my uncertainty about how students would process the experience, and hence signaled the need to facilitate more in-class reflections related to the activity. Indeed, this would be consistent with critical feminist pedagogy, which "demands sustained attention to the very epistemologies that underlie civic engagement discourse and projects as well as the pedagogical processes by which they are instantiated" (Costa and Leong 2012, 171).

Finally, the program is a host-stay program in which participants stayed with host families in Salvador. The experience of living with a family provides an unparalleled opportunity for students to develop their Portuguese and become immersed in Brazilian culture. However, the program currently only includes a small number of Afro-Brazilian families as host families. While there are some difficulties involved in locating middle-class black families who live along the shore in Salvador and who are interested and available for the program, we have begun to use the networks of past students from Instituto Cultural Steve Biko to locate such families.

Conclusion—Whose Global Citizenship?

Global engagement in Brazil is essential, especially given the country's growth and prominence on the global stage. It behooves students to have access to global opportunities in Brazil, and it is exciting to think about how summer programs might also be mutually beneficial for students in the host countries. Black researchers are uniquely positioned to develop exchange programs that target marginalized black populations in places such as Salvador. But, we are also located in a somewhat contradictory position when we do so. I not only encouraged students to study in Brazil, but also prepared them (especially black students) for the racial reality in Brazil. I have previously written about my own encounters with racism in Brazil, which included prejudice, stereotyping, and being profiled by the police (Hordge-Freeman 2015a, 2015b). This ambivalence was justified especially when, before we even arrived in Brazil, black students seemed to receive differential treatment in the Consulate office. In the first year of the program (in 2013), the visas of several black applicants were denied. Unfettered by constant Consulate rejections, one persistent black student drove four hours to the Consulate office—three times! On the last trip, she was accompanied by a USF abroad program officer, and her visa was issued. Certainly, the visa denials of students in the first cohort may have easily been a coincidence. But, in 2014, when we encountered similar issues (mainly with the visas of black students), where their passports were returned and more information was required of them than of other students, it seemed less like a coincidence. This experience and others represent the ways that global experiences and processes can engender a "new form of subjectivity," one that it is rooted in understanding racism in more global terms (Rizvi 2009, 260).

When I relayed these experiences with my students to other colleagues, some shook their head in disbelief and disappointment. I distinctly remember one colleague reassuring me that I would be relieved once my month-long stint in Brazil had ended and I could return to the United States. The underlying assumption was that in the United States racism was less of an issue, and I would be met with a different (and better) treatment. Indeed, at the end of the Brazil program, as I exited the plane in Miami and headed toward customs, it felt good to be home. The English that flowed over the intercom was reassuring, and as I prepared to pass through the last checkpoint with my belongings, I felt relief and flight exhaustion. At the checkpoint, my flight exhaustion was enough cause for me to be flagged as a suspicious person and I was directed to a private security check, where I was detained for over an hour. The security personnel informed me that because I was "acting delirious" (read: tired while black), they needed to check to see if I had taken and/or was trafficking drugs. Perhaps the only delirium that I had experienced was in that fleeting moment when I exited the plane and felt excited about being back home. Any semblance of refuge or relief that I might have felt about returning home to the United States slowly dissipated as I watched the security officer carelessly rifle through my belongings, half believing that I was a professor, and leaving no piece of luggage untouched. After over an hour of searching my belongings and questioning the veracity of my business cards, the bleached blonde allowed me to leave. She did not apologize but rather looked unfazed, if not perfectly pleased, that she had completed her job—a curious reaction, not the least because the blonde was also a dark-skinned black woman like me.

Earlier this year, I exchanged e-mails with a black Brazilian feminist from Salvador who appreciated my responses to her e-mails about collaborating on a project with a feminist group in Salvador. She offered me a disconcerting explanation for her gratitude saying that "*Muitas vezes pedimos ajuda e escutamos um imenso silêncio em troca. Sentimos que muitas irmãs americanas vêm ao Brasil fazer suas pesquisas sobre nós, lançar seus livros, mas, que não estão dispostas a nos ajudar a sair do lugar*" (Many times we ask for help but, in return, we hear an immense silence. We feel many of our American sisters come to Brazil to do their research about us, launch their books, but they are not willing to help us leave our place.) Her perception of the lack of mutual reciprocity between black Brazilian women and black US women researchers was devastating and indicting. My positionality as a black woman who has been the target of racism and sexism in both the United States and Brazil makes her point even more poignant.

For black diasporic scholars, our engagement with global communities may begin with a commitment to social justice, but the challenge is to establish or contribute to collaborative partnerships that can support these aims. In some ways, it has become increasingly easier to work with global community partners, especially in Salvador, as global currents mean that these spaces have increasingly come to adapt and expect the presence of foreigners (especially those from the United States). The inclusion of black Brazilian students into the USF in Brazil program is an important start in terms of demonstrating reciprocity and shared authority. As representatives of their own reality and as cultural ambassadors, their contributions allow for "subjugated knowledge" to be visible (Collins 1990, 18). But in addition to including reciprocity in programs in Brazil, Instituto Cultural Steve Biko has requested reciprocity in the form of more opportunities for black Brazilian students to study in the United States. In fact, Instituto Cultural Steve Biko has been extraordinarily successful at mobilizing diasporic connections so that several of its students have attended and graduated from Morehouse College in Atlanta, Georgia. Years later, US students from Morehouse College have also studied in Brazil, with a partnership with the Instituto Cultural Steve Biko, making this relationship a mutual exchange program. While I have incorporated and funded black Brazilian students as research assistants on projects in Brazil, a similar exchange involving black Brazilian students to the United States has not yet occurred at my own institution. However, I look forward to working toward this goal.

Long before the notion of global citizenship emerged, prominent black scholars such as W. E. B. Du Bois (who studied abroad in Germany) and noted scholar Anna Julia Cooper (who studied abroad in France) were keenly aware of the power of global engagement (Beck 1996; Evans 2009). Their trajectories illustrate that studying abroad can be a pathway to both global and political engagement. And if, as Kelley (2002) asserts, "revolutionary dreams erupt out of political engagement," then there is good reason to provide the next generation with opportunities that lead to sustainable global engagement (8). Those of us seeking solutions to persistent racism, sexism, and exploitation in Brazil and in other diasporic regions are faced with a formidable task. An approach that integrates teaching and research in ways that forge equitable, reciprocal, and sustainable global partnerships to complement the work already being done in diasporic communities is one part of a broader "emancipation methodological practice" (Hordge-Freeman et al. 2011). This practice also involves challenging rigid disciplinary norms, hierarchies of knowledge, and assumptions of research authority in order to recognize the expertise of local members of marginalized communities. Transnational researchers who are willing to integrate knowledge from local experts in their own teaching and research become allies in the fundamental struggle for what diasporic communities around the world have relentlessly demanded, and what noted feminist and historian Anna Julia Cooper calls "the very birthright of humanity": freedom.

Notes

1. USF professor, Bernd Reiter originally created the USF in Brazil program, and guided me as I expanded the program into Bahia, Brazil.
2. My first visit to Brazil, in 2008, was to Rio de Janeiro. Most of my other visits have been spent in Salvador, Bahia, Brazil.
3. This article from the *Chronicle of Education* illustrates how Harvard is prioritizing study abroad programs for its undergraduates and providing funds to support it. http://chronicle.com.ezproxy.lib.usf.edu/article/Harvard-Proposes-Overhaul-of/102329/
4. In Hordge-Freeman (2015a), I write about the way some black Brazilian researchers were skeptical of me because we studied similar areas, and they felt threatened because they believed that my work (on a similar topic) might automatically be viewed as more valuable because of my nationality.
5. In 2010, as I completed my dissertation at Duke University, I met Instituto Cultural Steve Biko member, Michel Chagas. A student in the Master of Public Policy program at Duke, our friendship led to useful critiques that have contributed to the development of the USF in Brazil study abroad program. He and his wife, Ecyla Chagas (PhD in Psychology) often host my visits to Brazil, and they have also advised me on more recent research projects in Brazil.
6. Evans (2009) offers an overview of the experiences of other prominent researchers in France who often note how they are treated differently based on having an American passport. She references W. E. B. Du Bois, Angela Davis, and James Baldwin among those who have written about these experiences.
7. Anonymous formal evaluations administered by the Department of Sociology averaged a perfect 5.0 score on all areas of evaluation of the program. Unfortunately, similar evaluation methods were not administered to Brazil partners.
8. Nancy de Souza e Silva (Dona Ceci) is a practitioner of Candomblé and was initiated into the Candomblé Temple of Ilê Axé Opô Aganju. She worked closely with Pierre Verger and is active in the Pierre Verger Cultural Center (part of the Fundação Pierre Verger) where she often leads storytelling classes about the African deities (Orixás) of Candomblé.

CHAPTER 4

Didn't Your Parents Like You?

Mojana Vargas

Introduction

Writing travel memories is not an easy task.[1] Organizing ideas and emotions is challenging because it makes us relive some unpleasant experiences. At the same time, it allows us to relive some moments that passed faster than we would have liked. This text is not the result of a thorough and systematic study. It is not intended to be a sort of guide for other students. It is an unpretentious attempt to share some of the experiences I have had in the past twenty-four months, which includes two different periods: a short stay in the United States, and the period after my move to Portugal to study for a doctorate in African Studies. This chapter is an attempt to bring together experiences of racialization that I lived or witnessed on these trips from the point of view of my daily life as a Brazilian immigrant[2] in a foreign country.

To begin with, some personal information may be useful in helping to contextualize the experiences that I will narrate. I am a black woman from a low-income, northeast Brazilian family.[3] From a very early age, I have been confronted with Brazil's deeply embedded social and racial hierarchies from my own family (Hordge-Freeman 2013). As the only black child in a family of white children, there was never room for any doubt about my racial identification. In effect, I was "the black" in the family group, and thus continued to be in all social domains I have been a part of ever since. Rather than my familial experiences being the catalyst for this reflection, the motivation behind this was the pressure I felt to continue my own educational process. In the beginning of 2013, I was quite anxious. Having acquired my master's degree in International Relations at São Paulo State University a few years ago and then having become part of the faculty of a public university, people began to ask: "When will you get a doctorate?" The answer was always the same: "Good question."

Back then, I had more questions than answers about pursuing a PhD, and I had not yet set a research topic or chosen a doctoral program. However, I could no longer postpone a decision because, for an academic career, the doctorate is essential to anyone who wants to pursue a major project. The need to have a PhD became a sort of professional obligation, or a formal step in the process of realizing one's "intellectual citizenship" on the path to developing meaningful research. Moreover, fulfillment of the PhD allows one to have access to the key element of current research: funding. In his book *Intellectual Citizenship and the Problem of Incarnation* (2013), Peter Eglin questions the meanings and consequences of being an academic in a major university and how this affects his capacity to keep a critical, but still socially and morally responsible, point of view in a contemporary capitalist world. He argues that any academic must question his or her ability to exercise intellectual citizenship without losing its integrity. My goal is much more modest. At this point in my academic career, I am still learning to assert my own *academic voice*, and the only possible path to this is through defending a doctoral dissertation. In this chapter, I will attempt to use my personal experiences to discuss how my racialization and struggles with challenging hierarchies of knowledge have impacted my academic work and my career trajectory.

Hard Times in Philadelphia

With the approach of the International Studies Association Conference, I decided to take advantage of the opportunity to present a paper that I had written and to explore potential future opportunities for pursuing a PhD in the United States, with its multitude of prominent universities. During this time in Philadelphia, I was asked several times to comment on the situation and treatment of black people in Brazil, because they assumed that race relations in Brazil were better. Their questions surprised me but reflected the way the idea of "racial democracy," which assumes the absence of modern racism based on Brazil's unique history, still shapes people's perceptions of Brazil (Freyre 2003 [1933]). More than people's questions, I was especially surprised given the racial composition of Philadelphia. The city has one of the largest concentrations of native black people in the United States, and a noticeable community of immigrants from West Africa and the Caribbean.[4] The city has also a rich history of rebellion against slavery and racism in which black women played a key role, as in the case of the Philadelphia Female Anti-Slavery Society.[5] In a way, I was frustrated with people's questions and disillusioned by what I thought I would find.

I was asked several times about this "harmony" between white and black people in Brazil and about living in a country where there is no racism.

The existence of a "racial paradise" in Brazil is still part of a widely held idea about the country, despite the amount of research that deconstructs this idea and shows just how much our race relations are based on historically constructed inequality and exclusion (Guimarães 1999, 2004; Telles 2003). I confess that, at times, I felt uncomfortable with the questions people asked about race. But what bothered me the most was that it seemed as though the academic debates about race and racism had not reached the people outside of the universities. During this time, I was not living with other academics. Living with nonacademic people with different levels of training made it very clear to me that we still have (myself included) difficulty in reaching an audience beyond those directly involved in race issues through academia or activism.

After two months, I decided not to pursue a doctorate in the United States for a number of personal reasons. Among them, of course, was the problem of language. In Brazil, students are given English classes during high school, but this activity is usually treated as something minor. As a consequence, students from public schools have this great drawback when they come to higher education. Moreover, the uncertainty about whether or not I could get a scholarship and the difficulty of adapting to a way of life so different from that which I was accustomed to were strong influential factors. At that time, I think I was perhaps more concerned with what it would be like to live in a country with a different mentality about how people should relate, with different consumption and eating habits different from my own, and with a problematic and narrow view about its place in the world. Being both a foreigner and black was a strange experience, especially because I noticed that people visibly altered their behavior around me when they realized I was from a different place. Most importantly, along with being defined as a foreigner, people would often talk to me about *Latino* culture, which was completely odd to me.

What Does a Brazilian Look Like? Racialization in the United States and Portugal

When I traveled to the United States, my concern was not so much on the question of race, but nationality. My expectation was that I would be seen generically as Latina (influenced by reports from other black Brazilian students who had been to the country), even among African Americans. When talking to people, my accent was immediately perceived as foreign, and when I identified myself as Brazilian, reactions varied according to the interlocutor. Some people responded to my Brazilianness by expecting me to sport a festive attitude that is supposedly shared by all Brazilians, a carryover from the association of Brazil with iconic cultural representations such as carnival, samba, bossa nova, or football (or soccer). Others—usually blacks—sought the image of Brazilian racial democracy (Freyre 2003 [1933]), as it was widely assumed that the absence of legal segregation of black people in Brazil equated to an absence of racism.

In Portugal, the situation was much more unusual. Soon after arriving in the country, I settled in at the student residence maintained by my university, so I could live with students from different backgrounds. During the reception for new residents, most of the attendants were non-Portuguese speakers, so everybody was speaking English at the party. Curiously, many people I talked to were surprised when I identified myself as Brazilian. By my accent, it was clear that I was not Portuguese (or even European), but my appearance was different from that which they associated with Brazil. One of the students even thought I was American. Days later, having become more familiar with these colleagues, I broached the subject, trying to understand their point of view. I wanted to know why they thought I did not look Brazilian. Could I be African? "No," they told me. Why not? Because compared to other African students, they saw me as "almost white." The other Brazilian students they had seen were white or brown with very straight hair, and I did not fit with those characteristics either. In their experience (as some of these colleagues have lived in the residence since the undergraduate program), they had seen dozens of Brazilian students, but I did not look like any of them.

Although unexpected, this observation was not entirely surprising. It was interesting to see how people of other nationalities perceive Brazilians. The fact that I was not perceived as a Brazilian at first glance reflects at least two important issues. First, it reveals the socioeconomic standard of the Brazilian students who regularly attend foreign educational institutions. Despite the changes introduced by government programs as Ciência sem Fronteiras,[6] the Brazilian students abroad typically have a white and middle-/upper-class profile, reflecting the majority of white students at this educational level (Feres and Zoninsein 2006; Nogueira et al. 2008). At the time I arrived in Portugal, ten out of about forty residents were Brazilians, and only one was black.[7]

The second issue highlighted by people's misconstruction of my nationality relates to the Brazilian image abroad. Portuguese media is very connected to Brazilian life, and people living in Portugal can enjoy one of the most successful Brazilian creations, the *telenovela* (soap opera). Nowadays, it is possible to watch five different Brazilian telenovelas airing at the same time, which is even more than what is shown in Brazil. Araújo (2000) notes that our telenovelas are characterized by a "Swedish aesthetic" and that the invisibility of blacks in the media has been used as part of the population-whitening process, which has been evolving since the end of slavery in 1888. Telenovelas have very limited black representation, and when there is a black character, they are limited to minor roles or they function as the villain. In this context, I understood that people's confusion over my nationality was connected to the mismatch they perceived between their images of Brazilians and blackness. Black people do not appear to be Brazilians because the Brazilians many people know or see are white.

This issue of hegemonic aesthetics was also at play with other experiences I had in both the United States and Brazil. In Philadelphia, during a subway ride, I noticed a couple staring at me. Initially I was concerned that there might be something wrong with my clothes or maybe something on my face. Although I was in doubt about whether or not to look back at them, when I did, the woman smiled at me and said: "I love your hair. Congratulations!" I smiled back and thanked her; after all, who doesn't like receiving compliments? However, I could not help but wonder about this situation. Why would somebody praise the hair of a stranger in the street in this way?

To my surprise, this happened a few more times and, intriguingly, it was always white people who passed me in the street or in the supermarket and fell in love with my "irresistible" African style. Similar incidents happened in Portugal when some people showed a fascination with my hair. I found it funny that some of them seemed unable to accept purely and simply that I had regular hair, like anyone else, only more curly and thick. And unlike people in the United States, the Portuguese seemed to have a more liberal concept of personal space. Some people could not resist the temptation of touching my hair to feel its texture, remarking "Ah! It's so fluffy! It feels like a sponge!" One can imagine my reaction to such statements. One particular colleague could not pass by me in the student residence without touching my hair and only stopped when I told her I did not like having my hair touched by anyone. The most interesting part of these encounters was the perplexed reaction of people when I asked them not to touch my hair. I seemed to be perceived as obnoxious for not wanting to be touched all the time.

In United States, I was met with criticism from a little African American girl who was clinging to her mother's hand while patiently waiting in line at a restaurant. She scowled at me and said: "You have ugly, nappy hair!" The girl might have been around four years old, too young to use chemical treatments for her hair. But her mother, a beautiful African American woman in her mid- to late-twenties wore her hair long and straightened (it also could have been a weave, but I am not sure). I found the interaction a little ironic. While white people seemed to feel the need to express their appreciation, black people simply ignored what they considered to be commonplace or sometimes offered a look of disapproval.

The issue of black aesthetics is complex because of the relationship between hair, identity, and self-esteem. As noted by Craig (2002), beauty standards have been used as a way to promote the domination of some groups over others, with the black phenotype being associated with ugliness and poverty. In the same vein, Figueiredo (2002) stresses the importance of hair to black women in Brazil, who uphold hair straightening as a norm for beauty and social mobility. In contrast, the same authors highlight the ways that black activists have used the black beauty pattern as a means of opposing racism by creating explicit links between black beauty and black identity, thereby affirming the value of African features and promoting the self-esteem of black people (Craig 2002; Hordge-Freeman 2015b). Both the United States and Brazil have a long history of regulating black bodies in such a way that white bodies become the norm. In this case, straight hair is used as a beauty standard to be achieved by all women, such that even white women try to keep their hair as straight as possible. During the rainy season, it is not uncommon to see black and white women trying to protect their heads from the rain using plastic bags—at once a ridiculous image and a testament to the efforts women undergo to maintain this beauty standard.

It is important to mention that the people involved in these interactions may certainly have very different perceptions about race, gender, and identity than what these comments suggest. The little girl's statement about "ugly, nappy hair" seems to express her internalization of ideas about black hair and beauty: you cannot be beautiful with curly hair. Although the debate on this issue has prompted better understanding and acceptance of natural hair,[8] the aesthetic pressure still affects women (and men) both personally and professionally. As a result, many men still opt for the shaved head solution. It seems that some white people feel compelled to help us raise our self-esteem by "helping us" to perceive and value our black beauty. They may also view these same "beautiful" styles as unprofessional in a work setting.

Portugal and Brazil: Two Faces of the Same Coin

Despite the fact that political separation between Brazil and Portugal occurred nearly two centuries ago, similarities between the two countries' social structures are still present, for better or for worse. Portugal, like Brazil, has a hierarchical society, in which titles and status are valued. The academic environment is special in this sense, since the forces that separate those with titles from those without add distinction to academics, resulting in legitimacy and, by extension, authority. Universities, like all spaces of power, are not neutral with respect to the asymmetries that exist in society, and they often reproduce or even reinforce many of them. Universities play a key role in both Brazil and Portugal as part of a national project in which they function as institutions interpreting major issues within a country and forming its elites (Trindade 2000).

Symptomatic of this custom are public events where discussions—when they exist—are preceded by long presentations in which all the works of the guests are listed and often commented on by coordinators of the panels, a reverence that at some moments becomes embarrassing. Of course, this is not meant to detract from those who contributed to the advancement of their respective fields, but it is interesting that it creates an atmosphere in which the audience attending the event is situated in an inferior position not only spatially, but also symbolically, in front of the parade of the guests' academic achievements.

There are other customs that emerge in our daily academic life that reflect these divisions. In Brazil, we have professors and researchers who insist on being called "professor doctors"; this distinction also exists in Portugal, including the use of a friendly nickname: *sotôr/sotôra*. When I first came to Portugal, I heard people using this term in reference to the dean of the institute where I studied, but I could not figure out what it meant. Of course, sometimes the term is used ironically; however, it is surprising how many people take it seriously. Once I was looking for a professor in his laboratory. I entered the room and asked the secretary, "Can I talk to professor John Doe?" And the secretary, looking at me somewhat impatiently, replied, "*Sotôr* John Doe is not coming today." I thanked her for the information and left, but I confess I only understood later what had occurred. Sotôr or sotôra is an abbreviated form of "Mr./Mrs. Professor Doctor," which is widely used in Portuguese universities in reference to professors,[9] especially the most experienced and renowned ones. She may have interpreted my calling a sotôr "just" professor as a sign of disrespect.

The use of these nicknames is just one example of several types of distinctions that are established in the universities. In Portugal, it begins with a process called the *praxe*, or the initiation of new students into academic life. This event reaches varying degrees in Portuguese universities, going further in those places where social life revolves mostly around the institution, such as the University of Coimbra. The *Academic Praxe* is the term used to describe the set of rules that should guide student life, and its cornerstone is the hierarchy established between the older students and freshmen, so that the latter must submit fully to the authority of the former, the so-called senior students (Dias and Sá 2014). Heavily criticized in Portugal, the *praxe* is a tradition among students who see it as a rite of passage necessary for the entrant, a way to prepare new students for the challenges they will encounter in adult life, which, from the perspective of *praxistas*, will be much worse than the challenges faced during their time in university. Besides the process of subordination to which freshmen are subjected, there are common occurrences of recklessness or physical and psychological violence[10] such as intimidation and even social exclusion of those who refuse to submit. Additionally, the praxe involves an important aspect: it reinforces the trend in universities to replace critical thinking and a quest for innovation with conservative and standardized thinking (Ferreira and Moutinho 2007).

How, then, does the *praxe* affect Brazilian students? It has a significant impact on them, since Brazilian students who enroll in Portuguese universities through federal programs have one more element to contend with in the *praxe*; in addition to being freshmen, they are foreigners. More importantly, some are blacks, making them even more vulnerable since the codes of praxe (hazing) allow the use of all types of derogatory comments. As a graduate student, I do not have to submit myself to *praxe* but it is very striking to see freshmen circling the university carrying signs that designate them as beasts and singing racist, sexist, xenophobic, and homophobic songs. It makes you think about how hierarchies are built in the institution and how it affects interpersonal relationships. When observing the initiation rites of students, critics usually give little consideration to the violence used in the process and, in general, little is said about the motivations of senior students, particularly the racist motivations that guide their actions (Ferreira and Moutinho 2007).

I believe that the lack of commentary about the racial elements of the praxe stems from the idea that racial democracy is still strong among the Portuguese. In Portugal, as in Brazil, there is a widespread view that racism does not exist as something structural in social relations, but only as an isolated act of ignorant people and as an attitude that is always assigned to the other. Even when they are included in the institution, the groups of Indians, Chinese, Gypsies, Africans of various origins, or black Portuguese end up being excluded again in a different way. It is after they join the Portuguese universities that the students belonging to these minority groups are confronted with the fact that there is no place for cultural diversity (Chinese, Indians, and Pakistanis are simply "Asiatic people") and even less space for the *knowledges* produced by the communities from which they originate (Santos and Almeida Filho 2008). After all, their contributions are not considered scientific knowledge.

This same *tábula rasa* reproduces the hierarchical structure and rules that classify different fields of knowledge. According to Andrioli (2005), this framework is perpetuated by the university itself when it establishes the division between the *hard sciences* (the true scientific knowledge, based on controlled experimentation), the *social sciences* (the environment in which scientific knowledge is almost validated as long as the research is conducted according to the methods of the hard sciences), and the *humanities* (areas of knowledge in which science is replaced by the narrative).

In the hierarchy of knowledge, the area of African Studies is certainly at a disadvantage, at least with regard to the institution that I attend. Recently, there was a call from the Science and Technology Foundation for funding of research projects and, as always, there was a veritable gold rush across the institution. At the Center for African Studies,[11] recognized as a unit with considerable intellectual production, there was the expectation of getting enough funding to pay for all projects submitted. But, to everyone's surprise, all requests were denied. There were many different explanations, but one factor is obvious: although it is a center of studies on Africa, no African researchers are linked to the entity. There are very few black students, as well, including Portuguese ones. There are master's and doctoral students of African origin, but all the professors that comprise the center's faculty and the scientific council of the institution are European (most are Portuguese, but there are also French and German scholars), and this is a taboo topic in the department. Why are there no African professors in the ISCTE/IUL Center of African Studies? According to the Dean of Coimbra University who has given several speeches about how Portuguese universities are suffering from public budget cuts, currently, the major problem is the economic crisis affecting all institutions of higher learning since 2009. However, before now, has there not been a single African researcher with the qualifications necessary to join the faculty of the institution? Note that the preparation or competence of current faculty who belong to that group of scholars is not in question; on the contrary, all have shown great experience and true expertise in their respective fields of research. However, even to maintain the coherence of an institution that wishes to be pluralistic, the African presence in the Center for African Studies is essential.

If, in the official plan, the discourse of integration and interculturalism[12] prevails, the relationships are a bit more problematic in practice. Unlike in Brazil, blacks are actually a numeric minority in Portugal,[13] and, as the item "race" is not included in the Portuguese census, there are no organized data about this group in the country; they are practically invisible, as pointed out by Araújo and Maeso:

> ACIDI's (High Comissioner for Immigration and Intercultural Dialogue) speech and political practice emphasizes a conception of racism and xenophobia as the derivation of a poor integration of immigrant communities and ethnic minorities. Its own sphere of activity can be seen as a kind of pendulum movement between the need for knowledge of the "Other" from the majority society and to favor an active integration in the "native" society, especially in the economic and cultural spheres. In this framework, anti-racist approach is not considered a priority. It is the successful integration (usually read as assimilation) of immigrant communities and minorities—considered more vulnerable against racial discrimination—that is seen as the natural antidote to racism. (2013, 152)

According to the last census, Portugal has 10,562,178 inhabitants,[14] including 394,496 foreigners with permanent residence. According to a survey based on ACIDI records, from these, 113,159 are immigrants with permanent visas from Cape Verde, Angola, Guiné-Bissau, São Tomé and Príncipe, and Mozambique, all former Portuguese colonies. In other words, African descendants comprise approximately 1 percent of the Portuguese population (Santos et al. 2012). But, since the beginning of decolonization, African immigration to Portugal has increased in a continued pattern, and despite a lack of official statistics, it is possible to find estimates of around 500,000 black Portuguese, including legal African immigrants and Portuguese with African descent. However, the number may be even greater because of illegal immigration. Moreover, problems in these cases are also more complex, as in the representative case that affects the community of immigrants from Sao Tome and Principe. Most of this community consists of women who survive as domestic servants and face serious abuses from employers. However, most of the complaints do not reach the judiciary branch because almost all of these workers live illegally in Portugal. Fearing deportation, they eventually end up submitting to the precarious conditions of their work.

Immigrant populations are undergoing a process of constant segregation. While populations of African origin—the so-called historical immigrants—occupy mostly the outskirts of Lisbon, the town itself is predominantly occupied by groups of more recent immigrants, such as Indians, Chinese, and Pakistanis, who are confined to run-down areas, as is the case of the Mouraria neighborhood. Nevertheless, the fact is that while everyone intermixes and circles the city during the day, at night there is a noticeable presence of people of African descent across the Tagus river toward the dormitory towns[15] (Malheiros and Vala 2004; Faria 2010).

Research conducted by nongovernmental organizations and research institutes shows that the black population is especially vulnerable in Portugal. Black people are underrepresented in political and institutional decision-making processes and have greater difficulty in accessing education (especially at the college level) as well as public services and employment. Malheiros and Vala (2004) show that immigrants from Portuguese-speaking countries in Portugal represent just 3.8 percent of the number of immigrants with university degrees (1074–1077). Since they are less educated, they suffer more with unemployment (11.8 percent) and precarious housing conditions; approximately 40 percent of them live in slums or very dilapidated buildings. Although the country has legislation criminalizing racial discrimination, official bodies do not keep specific statistics on various racial and ethnic groups that would serve as a way to monitor the vulnerability of each, arguing that adopting specific policies for each minority could create divisions and animosity where none exist; it would also challenge the official policy of assimilation (Freire 2007; Reiter 2012). Again, the similarity with Brazil is no coincidence, though in this aspect, thankfully, Brazil is more advanced because of the pressure from organized social groups.

The so-called intercultural politics are not exactly a novelty in Portugal because their roots go back to the period of the Salazar dictatorship,[16] whose discourse of racial integration was put forth with the objective of assisting in the maintenance of the Portuguese colonial empire in Africa (Vala et al. 2008). This discourse sought to show the world that unlike the northern European settlers, the Lusitanians in Africa did not mistreat or discriminate against black Africans in territories under their control. The Luso-tropicalism inspired by Gilberto Freyre's work was very effective at concealing racism and its structural subordination mechanisms to prevent debate on the issue. On more than one occasion I had the opportunity to discuss this with people who simply deny the existence of racism in Portugal, as well as the existence of discrimination against blacks (African or Portuguese of African descent), Asians (who are very numerous in Portugal, especially Indians and Pakistanis), or gypsies. Moreover, there is a tendency to blame victims of racism for causing racism.

However, it is in the day-to-day life that one can realize the most obvious similarities between Brazil and Portugal when it comes to racism. In Portugal, as in Brazil, relationships are not openly hostile; on the contrary, friendliness is a key characteristic of the way people relate to each other. The different forms of racial prejudice and discrimination are characterized mainly by subtlety and cordiality (Marques 2007).

Are You Sure It Wasn't Just a Joke?

During my first days in the university, I witnessed an unusual situation in the copying services department. I needed to finish the registration of my records in the institution, so I spent much of my time copying or printing documents to deliver to the Academic Office. In general, the copier attendant was always having fun with students who apparently had no problem participating in his jokes. However, on one occasion he ended up having a small argument with a student named Jonathan. I don't know why the clerk asked the name of the student, but when the student answered, the employee adopted a humorous expression and fired back: "Jonathan? What kind of name is that? Didn't your parents like you?" The boy, visibly annoyed, replied: "I don't know, but that's my name, so it doesn't matter." The boy quickly gathered his stuff and left, while the employee continued to comment on the situation without understanding the boy's annoyance. After all, the employee was not guilty of the strange name given to the boy,[17] and furthermore, it was just a joke. He could not understand why these foreigners gave such strange names to their children. Ironically, that attendant was a mestizo, son of an Indian and a Portuguese, according to what he told another student.

I could not believe what I saw and heard. The clerk had just openly offended an African student because of his name. Unbelievable! I replayed this scene and everything that occurred before it in my mind, in order to ensure that I had not missed anything, and left. Days later, I returned to the same place to scan some documents. I handed a business card to the attendant—the same one as before—and asked him to send the files to my e-mail address. He read my card, grimaced, and said: "Mojana? Didn't your parents like you?" Again, I was taken aback with disbelief. Not only because this embarrassing situation was happening with me, but also because of the poor judgment the attendant used in repeating a joke he knew was not well received. Moreover, the individual seemed to feel pleased to have embarrassed others—not just the targets of his pranks, but also some of the silent witnesses. But not today. I took a deep breath and replied: "On the contrary. They liked me so much they gave me a name full of personality!" He got the message. He then sent the e-mail, handed me the card, and thanked me. A young woman, who until then was looking at me with a little pity, looked at the clerk and started laughing, perhaps herself a former victim of his "jokes".

In Portugal, as in my country, some individuals tend to minimize racial slurs, sometimes justifying them as failures of communication among the individuals involved or even as reverse racism. When I reported the case to my fellow residents, some of them told me that, in fact, everything should have been taken just as a joke and that the racism I perceived was in my own head. Unfortunately, this kind of view is widespread in many places involving the black diaspora. Sue and Golash-Boza (2013) provide a broad outlook of how this subject is dealt with in many Latin American countries, especially in those places where "colour-blindness" is a central part of national ideology, and "created a social silence on the topics of race and racism outside the humorous contexts" (1584). They also show how humor is related to the reproduction and naturalization of inequalities between social groups. Academia has not paid much attention to this subject in Portugal, but Martins (2012) provides an initial description of what he calls "ethnic humour" between Portuguese people (90). Despite the apparent innocence of the phrase, it is clear how this *ethnic humour* might direct prejudice against its target.[18]

Speaking of jokes, racism is one of the most frequently used sources of material by national comedians. In Brazil, organizations of the black movement have increasingly reported racist jokes made by Brazilian comedians who, unfortunately, have incorporated a discourse of exclusion and dehumanization of their targets, whether they are homosexuals, women, blacks, or the poor. To defend their positions, comedians accuse critics of imposing excessive surveillance and even censorship in the name of an alleged regime of "political correctness."

Likewise, a joke from Portuguese comedians have been used to internal dispute about the new direction of the Portuguese Socialist Party. The parliamentarian António José Seguro, a white, traditional politician, former Deputy Prime Minister of Portugal, is opposed to António Costa, also a Portuguese parliamentary, a mestizo born in Goa[19] and Mayor of Lisbon. As a type of word game, a columnist of the daily newspaper *Jornal de Notícias*[20] launched the joke: "Don't vote in the dark, vote Seguro."[21] The column had a considerable number of readers who approved the line of reasoning of the author, who, by the way, was not even worried about responding to the several criticisms of racism received.

Despite the project of interculturalism and the subversion of racism provided by common practices of subtlety, tension is everywhere: in business, in public transportation, and in the streets. It has only been a few months since I came to Portugal and I have been through many situations that are quite familiar to Brazilian black people, from being followed by security guards in stores to being refused by taxi drivers. A security guard followed me closely at a health food store while I was choosing granola as, clearly, a black person in the cereal section is largely out of place. Likewise, the university community also reproduces these conflicting relationships, albeit in a subtler way. One of the most common situations is when I introduce myself as a PhD student during a conversation and the other person asks me if I really want to move back to my country after graduation. I mean, what kind of question is that? Personally, I have no intention of emigrating, but what if I had? Would that be a problem?

When I decided to join the PhD program in African Studies, I decided to take a French course offered by the university, so I would be able to read the huge production of information in French language in this area. Unfortunately, I had to start from scratch and, in one of the first sessions, the teacher was teaching us the names of colors. At one point, a classmate who sat next to me asked how to say "skin-colored pencil". The teacher was somewhat perplexed by the question. The girl insisted: "Skin-colored, teacher!" And while speaking, she was pointing to her own arm. The teacher replied: "Skin-colored? What color is that?" The girl continued pointing to her own arm: "Like this. A normal skin color!" So I asked: "What do you mean *normal* skin color? What would be an *abnormal* skin color?" The girl looked at me perplexed and did not say anything. The teacher thought it was better to drop the subject. For the next class, the girl sat elsewhere.

I decided not take up the issue to avoid further problems, but I am curious to know what goes through the mind of someone who says something like that. I wonder if she realizes the absurdity of her words or if she really thinks that whiteness is something normal, or even the *only* race to be considered normal. I venture to say that my former French class colleague shares the opinion of many Brazilian and Portuguese people who consider racism as existing only among black people, as if black people will see discrimination in anything that is said or done by white people. After all, Brazil is a racial democracy and Portugal is a country of "mild manners," where people have black friends and, supposedly, skin color does not even matter.

Conclusion

Brazil and Portugal have an official discourse based on a supposed integration of diverse people. In the same way that the social sciences have deconstructed the discourse of racial democracy in Brazil, nowadays, the increase in African immigration has brought a new challenge that the discourse of cultural diversity has swept under the rug of interculturalism: racism is invisible in this society.

Even though racial relations is not my field of expertise, as a black woman I cannot avoid the effects of race-based exclusion and oppression that exist in Brazil, Portugal, or the United States.

From the starting point until the end of the first stage of academic life, this exclusion is present. Beginning with getting accepted to college, one starts to be classified, trained, and included/excluded from groups according to one's social and racial background. Students in Portugal, Brazil, and the United States are trained to accept certain *knowledges* as valid and to dismiss others. In fact, one's successes and challenges are sometimes linked to how well one is able to internalize these ideas (Hordge-Freeman et al. 2011). As an academic, I feel responsible for thinking about these kinds of subjects, inasmuch as they affect the whole social context in which I am involved, personally and professionally. I think this is also part of the process of acquiring "intellectual citizenship"—not only the formal academic process, but also becoming capable of organizing observations and ideas in a way that they can be integrated into your research agenda, contributing to the advancement of one's knowledge and social equality.

It is important to note that the voice that comes with "intellectual citizenship" is not exclusively one's own. My current research is about Brazilian Policy for Africa and what inspired me was the need to develop a *black perspective* about black people in Brazil, Africa, and around the world that could be used in my academic context, considering that this standpoint has still limited representation at universities.

In spite of what you try to convey, some of those in academia will try to discipline you in the same way as the senior *praxe* students do to freshmen. And others will respond differently. The next stage in this process for me is to learn how to deal with these challenges.

Notes

1. As some of the examples were written from memory and sparse text notes, the reader may note a lack of detail in some places. I supplement this with the analysis of my experiences. I am very grateful to Gladys Mitchell-Walthour, Elizabeth Hordge-Freeman and Ana Rita Alves for their comments and suggestions on the bibliography.

2. Ordinarily, Brazilian college students and researchers in Portugal do not see themselves as immigrants because they consider their stay in Portugal as temporary (Sardinha 2009). This changes only if or when they begin to consider the possibility of not returning to Brazil. However, in the eyes of Portuguese people and experts in the social sciences, we are immigrants, although some may see us as an "improved" category of immigrants, in what is described by Machado (2006, 120) as "hierarchy of alterity."

3. The migration of families from the north and northeast to the southeast of Brazil was especially intense between 1950 and 1980, owing to ecological or socioeconomic circumstances (see Póvoa Neto 1990).

4. US Census Bureau. State & County Quick Facts. http://quickfacts.census.gov/qfd/states/42/4260000.html

5. Our Sphere of Influence: Women Activists and the Philadelphia Female Anti-Slavery Society. http://hsp.org/sites/default/files/legacy_files/migrated/pfassstudentreading.pdf

6. Study abroad programs maintained by the Brazilian federal government give funding to undergraduate and graduate students attending public institutions. Previously, the student or his/her family would pay all costs. Unfortunately, the item specifying race/color of the participants is not available in the data about the scholarship distribution. See: http://www.cienciasemfronteiras.gov.br/web/csf/painel-de-controle

7. According to the Brazilian Embassy, there are 9000 Brazilian students in Portugal (both graduate and undergraduate), but there are no available data about the distribution of race and color.

8. "Natural hair" here is used to mean not modified with chemical products, straightened with flat iron (the popular *chapinha*), or manipulated with extensions or weaves.

9. In Portuguese, the word "professor" is used to describe anyone teaching at a school, institute, college, or university. In Portugal and Brazil, most university professors are in the same position as tenured professors in the United States.

10. In December 2013, seven students of the Lusophone University of Lisbon were swept into the sea by a large wave during an event of academic *praxe*. Only one

of them survived and after a confidential police investigation, the Portuguese courts decided to close the case as it found no evidence to blame the survivor or anyone else that could be involved. In April 2014, three students of Informatics—girls aged between 18 and 21 years from the University of Minho—died after being hit by a falling wall during an academic *praxe*.

11. By the time I was working on this chapter, the former Center for African Studies was closed and the researchers linked to it were absorbed by the new Center for International Studies. Part of this change is related to the cut on the national budget, but the decision is also based on the lack of interest in this field.

12. The High Commissioner for Immigration and Intercultural Dialogue (ACIDI, the Portuguese government agency responsible for the reception and integration of immigrants and ethnic minorities such as gypsies) defines interculturalism as a project of transforming a social group formed from various origins—a multicultural group—into a cluster in which different cultures intersect in a process of mutual transformation.

13. In Brazil, the black population (black and multiracial) is estimated to be 51 percent of the Brazilian population. http://www.sae.gov.br/site/?p=11130

14. Instituto Nacional de Estatística: Estatísticas Oficiais: As Pessoas (2012).

15. "Dormitory towns" are the cities around Lisbon that integrate the Lisbon Metropolitan Area, where most of the population lives. Housing in Lisbon has turned progressively more expensive each year, especially since 2008. Most people who work in Lisbon live in cities like Sintra, Oeiras, Odivelas, and on the south bank of the Tagus River, where the rent is cheaper.

16. Between 1932 and 1968, Portugal was governed by António de Oliveira Salazar, who took office after a coup led by the Portuguese military in the previous year. The new state was a fascist-inspired regime in which the political police (PIDE), the Catholic Church, and the nationalist discourse played a key role. As a way to keep the Portuguese colonial empire, Salazar adopted a foreign policy of economic isolation and political neutrality in the face of the international powers of that time.

17. In Portugal, only words considered typically Portuguese are accepted as surnames in birth certificates. See more at http://www.irn.mj.pt/IRN/sections/irn/a_registral/registos-centrais/docs-da-nacionalidade/vocabulos-admitidos-e/

18. He describes a popular joke from the Liberation of Colonies War era: "Do you know when a nigger reaches a higher level in his life?" [No] "After stepping on a landmine."

19. Goa is a region in India that integrated the Portuguese colonial empire between 1510 and 1961, when it was taken over by the Indian army.

20. http://www.jn.pt/Opiniao/default.aspx?content_id=4009903&opiniao=Jorge+Fiel. Published July 5, 2014.

21. I know that explaining jokes is always a bad idea, but in this context, *dark* refers to Costa's skin color, and *Seguro*, in Portuguese, means *safe*.

PART II

RWB: Researching While Black and Female in Brazil

A (Black) American Trapped in a ("Nonblack") Brazilian Body: Reflections on Navigating Multiple Identities in International Fieldwork

Tiffany D. Joseph

Introduction

Though qualitative researchers have examined the influence of their position-alities on their research in the United States, less is known about how such positionalities play out when conducting international race research. From October 2007 to October 2008, I conducted fieldwork in Governador Valadares (GV), Brazil, a small city in Minas Gerais, to examine the racial conceptions of 49 Brazilians who migrated to the United States and subsequently returned to Brazil. I aimed to learn how migration influenced these individuals' understanding of racial classification, stratification, and relations in both countries. Furthermore, as GV has historically been Brazil's largest emigrant-sending city to the United States, I wanted to explore if this extensive emigration and return migration to GV had also altered race relations there.[1]

As a qualitative sociologist, I was well aware that being a lighter-skinned black, middle-class, and educated American woman would have some impact on how I interpreted Brazilian race relations, and how my Brazilian respon-dents would racially classify and interact with me.[2] Historically, it has been presumed that race is an individual's skin tone and that higher social status can yield social whitening for nonwhites in Brazil; this is especially the case for high-status black women who are rarely classified as black by other Brazilians (Degler 1986; Telles 2004; Daniel 2006; Schwartzman 2007; Bailey 2009). Thus, I found that Brazilians I interviewed and encountered throughout my fieldwork did not see me as a black because of my skin tone or as an American because I was not white. This chapter examines my reflections conducting this research as a self-identified black American woman who was not always perceived as such in Brazil. Before delving into these reflections, I briefly highlight some relevant scholarship on researcher positionality, and explain how I established rapport in the field site and acquired an understanding of Brazilian race relations to conduct my research. Next, I delve into how my appearance influenced Brazilians' perceptions of my nationality and racial classification, which shaped their interactions with me. I conclude with a discussion of how engaging with people in the field changed me, more specifically my perceptions of the different dynamics surrounding skin tone in Brazil and the United States. In doing so, I demonstrate that, as researchers, who we are and how we look influence our fieldwork whether we want it to or not. However, by being aware of how our positionalities influence research, we can use them to connect with our respondents and gain significant insight into their perceptions of and experiences in the social world.

Theoretical Background: Positionality and Reflexivity in Qualitative Research

The role of researcher positionality and objectivity has long been debated in social scientific and qualitative research because researchers, like respondents, have multiple and intersecting social positions that influence their perceptions of the world and how they conduct and interpret research (Bourdieu and Wacquant 1992; Collins 1998; De Andrade 2000; Dwyer and Buckle 2009; Mazzei and O'Brien 2009). Consequently, qualitative researchers have developed the concept of reflexivity to reflect on their engagement with positionality in the field (Collins 1998; Doucet 2008; Dwyer and Buckle 2009). Researchers can inhabit "insider," "outsider," or "insider and outsider" positions (Dwyer and Buckle 2009). While an "insider" shares at least one type of status group membership (e.g., race, gender) with participants, an "outsider" has no membership with participants. Nevertheless, it is possible for a researcher to simultaneously possess "insider" and "outsider" status since researchers adopt multiple social group memberships. Sociologist Patricia Hill Collins (1998) refers to this multiple positionality as "the outsider within," which describes individuals who find themselves in marginal locations between groups of varying power. Sociologist Andrea Doucet (2008) suggests that researchers can better understand reflexivity in their research through what she calls three "gossamer walls," which influence the relationships between the researcher and (1) his/her multiple selves, (2) respondents, and (3) audiences who will read and engage with the work.

However, the majority of these reflexivity concepts focus on ethnographic studies conducted in the United States. There is very limited inquiry regarding the role of researchers' reflexivity and that of respondents when conducting international research, especially for US citizens doing fieldwork abroad (Ansell 2001; Chong 2008). US researchers symbolically carry their nationality into the research site, which means they become informal representatives of the global social, economic, and political power of the United States. This, beyond race, class, and gender, can clearly influence researcher–participant interactions by introducing additional power and social dynamics. It is especially important for researchers to be aware of, understand, and minimize this power in order to ground the data in the personal experiences of participants (Mullings 1999; Hunter 2006). This requires researchers to be knowledgeable and conscious of participants' social context and the research site.

All of these considerations were especially important in my project since I was conducting cross-lingual and international research on race with people whose first language, nationality, conceptions of race, and socioeconomic status (in some cases) were very different from mine. I had to establish insider ties within the community so local residents, known as Valadarenses, would feel comfortable sharing their perceptions with an outsider from the United States. My Portuguese proficiency and preexisting contacts with local immigration researchers provided some legitimacy and helped me establish rapport with Valadarenses. However, being a temporary migrant also made participants feel at ease during interviews. Although my reasons for going to GV were different from the motivations my Brazilian respondents had for migrating to the United States, we shared the experience of voluntarily leaving our homelands to live in another country for an extended period of time. Respondents and other Valadarenses were impressed by my willingness to learn Portuguese, leave the United States, and live in GV to conduct fieldwork.

Perhaps most importantly for the project, I had to bridge my understanding of race as a black American sociologist with that of my Brazilian participants in order to learn how race functioned in GV specifically and Brazil more broadly. Though I was already familiar with existing scholarship on race in Brazil, it did not completely help me comprehend Brazilians' everyday perceptions of how race functioned in their country. This was especially the case because most of those studies had been conducted in large Brazilian cities like Rio de Janeiro, São Paulo, and Salvador, and none had been conducted in GV. Furthermore, racial demographics vary depending on where one is situated in Brazil. While northeastern Brazil has a large black population due to the importation of enslaved Africans in the seventeenth century, southern Brazil has a large white population due to significant European migration after the two world wars (Telles 2004). Much of the country's indigenous population resides in western Brazil (Telles 2004). However, in GV and Minas Gerais, most individuals have physical features that fall in the middle of the black–white continuum and are characteristic of the Brazilian phenotype: light- to medium-brown skin and straight black hair. For this reason, 56 percent of Valadarenses and 44 percent of Mineiros (residents of Minas Gerais) compared to 43 percent of Brazilians nationally self-classify as pardo (IBGE 2010). In my conversations with Valadarenses, I learned that many of them perceived GV as more racially mixed and less racist than other parts of Brazil (Joseph 2015). Consequently, Valadarenses interpreted their own racially mixed phenotypes as an embodiment of the national racial democracy ideology (Joseph 2015). Thus, Brazilians use local and national level discourses of race to interpret dynamics in different parts of the country. Living in GV and traveling throughout Brazil allowed me to be socially and culturally immersed in the culture and to acquire the local and national nuances of Brazilian race relations and how they differed from the United States. My formal and informal conversations with Brazilians, reading newspapers, and watching Brazilian movies, telenovelas (soap operas), and television programs were also essential for this immersion.

Conflicting Positionalities: Race, Skin Tone, and Nationality in the Field

Like my respondents, I have multiple social identities that are negotiated very differently in the United States compared to Brazil. As a native Southerner and descendant of enslaved peoples in the United States, I ethnically identify as African American. Racially, I identify as black though I have multiracial ancestry. While in the United States, one drop of black blood has primarily been the basis for ascribing black racial group membership to individuals, having multiracial ancestry in Brazil allows individuals to not classify solely as black, but as brown or even white in some cases (Degler 1986; Davis 1991; Marx 1998; Feagin 2000; Telles 2004). Thus, even before conducting my fieldwork, I knew it was likely I might not be considered black in Brazil. After all, my racial classification had often been questioned in the United States and abroad in my previous travels. Because of my light brown skin and the shape of my eyes, lips, and nose, I have been mistaken for Dominican, Puerto Rican, Brazilian, and biracial (black and white, black and Asian), among other ethnicities depending on the language I spoke, how my hair was styled (straight versus curly), and where I was geographically.[3]

Upon my arrival to GV, I learned quickly that if I did not say anything while going about my daily life, no one would suspect I was not Brazilian.

However in some cases, the moment I spoke in foreign-accented Portuguese, Valadarenses often assumed I was from another part of the country since accents vary throughout Brazil, or that I was born abroad to Brazilian parents. Because I wore my hair in a natural, very curly style, many Brazilians thought I was a "Carioca" from Rio de Janeiro or a "Baiana" from Bahia, both of which are regions that have large black and pardo populations in Brazil. My "Brazilian" appearance gave me the ability to blend in and provided pseudo "insider" status, which was beneficial for my research and provided entrée into locales that might have otherwise been inaccessible.

However, upon disclosing my US nationality, shocked Brazilians were quick to say: "Wow, I thought all Americans were white" or "You don't look American, you look Brazilian." I found that Brazilians did not believe I was from the United States because I was not white. As a consequence, I often had to "prove" my Americanness by explaining my background as a multi-generational US citizen whose ancestors had been in the country for centuries. I often had to explain my family background, telling Brazilians that I am a descendant of enslaved people in the United States, slave-owners, and Native Americans.[4] After these long explanations, Brazilians finally believed I was "American."

Within the specific context of GV where I lived, Valadarenses often assumed I had migrated to the United States and had naturalized, or that I was born in the United States to Brazilian or other Latin American parents. I acknowledge that this may have been due to the extensive GV–US migration history, which has created significant transnational families in the city. Nevertheless, some study participants blatantly told me when I arrived to conduct interviews that "I wasn't expecting someone so Brazilian looking, but someone with blond hair and blue eyes."[5] Although I did not visibly flinch or change my body language in response to such remarks, those words evoked mixed feelings within me as a researcher.

On the one hand, I think people were complimenting me to signal that they saw me as a Brazilian "insider" since my phenotype resembled theirs, which put people at ease when sharing their experiences. However, such remarks also demonstrated how participants attached ideals of race, skin color, and nationality to me within the interview encounter. In viewing my nonwhiteness as not authentically "American," participants and other Valadarenses I encountered recognized that some phenotypes were "more" American or Brazilian. Thus, my appearance and Portuguese proficiency challenged Brazilians' preconceived notions of who was American or Brazilian, which then influenced their perceptions of and interactions with me.

How Valadarenses and I Racially Classified Each Other

Although Valadarenses thought I looked Brazilian, my data indicated that my respondents saw me as a nonwhite Brazilian: the majority classified me as pardo or "other."[6] Given the perception that GV is very racially mixed, most Valadarenses likely classified me as nonblack since my skin tone was similar to theirs. The quotes below demonstrate some participants' responses when I asked how they racially classified me.[7]

> You, for example, you are a mulata or a parda [brown] woman because you are neither black nor very dark . . . because of your appearance, your [skin] color.
>
> Portuguese man, aged 31

> You? Here in Brazil or there in America? In America, you would be black, here you are morena. My daughters are brown [too].
>
> White woman, aged 36

> You are morena here in Brazil, not black. You should have observed this in other places. So, I am considered white and a person of your color is morena.
>
> White man, aged 50

> I think you are mixed, a mix of black and white.
>
> Black woman, aged 31

> [I classify you as] Other. You are not white and you are not totally black. You are not parda [brown], yellow, nor Indigenous. (*laughs*) So, I do not have words [to classify you] because I do not think much about this thing of color. For me it is not important.
>
> Brown woman, aged 41

Notice how in each of these quotes, respondents told me I was not white or not (completely) black, but rather "mixed." Since each of these respondents had also lived in the United States, a few asked if I wanted them to classify me based on US or Brazilian standards and subsequently responded that I was black in the United States, but not black in Brazil. It was through such exchanges that I was able to reconcile that I was a black American trapped in a nonblack Brazilian body. Furthermore, when I asked for participants' racial conceptions, they referenced my physical attributes to clarify a point they had made or to describe their own racialized phenotype in relation to mine. The quote below is one example and also demonstrates how participants ascribed "Brazilianness" to my physical appearance:

> If you don't say you are American, no one would say you are American in Brazil. Folks have already told you this huh? [In terms of] hair, skin color, face, and everything, if you don't open your mouth [speak], [people] will think you are Brazilian. . . . In Brazil, you are black. Black, no, you are mulata,

a mixture of black with white. You are considered what we call morena, coffee with milk. . . . Your color, it is not white nor dark. (*laughs*) You are the color of Brazil.

Mixed woman, aged 48

Just as I asked participants how they racially classified me, I also collected data on how I would racially classify participants using Brazilian Census categories and their physical attributes. This allowed me to incorporate a "Brazilian" racial lens and minimize the influence of my US norms for racial classification when classifying participants. I generally classified individuals with much lighter skin and straight hair as white and individuals with very dark skin and "curlier" hair as black. I had more difficulty classifying individuals whose phenotypes were in between that range, for example, from light to medium skin, dark loosely curled hair, and light eyes. In such moments, I considered that individual's attributes compared to my own. For example, if their skin tone was similar to mine, I classified them as pardo since most Brazilians considered me to be pardo. If their skin tone was significantly lighter or darker than mine, I classified the individual as white or black, respectively.

Even using this method to classify my participants was not always reliable, as I learned I could not always expect participants to racially classify in a manner in which I expected. In one case, I interviewed a female returnee whose skin complexion was the same as mine and she self-classified as white. When I asked her how she classified me, she said: "Well I see myself as white and since you are my color more or less, I would say you are white too." Even though I recognize that white in Brazil and white in the United States are very different classifications, in the Brazilian context, I was baffled as to how this young woman with my skin tone could have self-classified as white. However, the longer I stayed in GV and learned the norms of racial classification there, people's assessment of their own and my classification surprised me less. I found that I had begun to see race, or rather skin tone, with a more Brazilian lens.

Receiving Differential Social Treatment as a Nonwhite American and Brazilian

As I became personally immersed in the nuances of race in GV and Brazil, I also recognized that I was treated differently when people assumed I was Brazilian and subsequently learned I was American. Although it is possible that my being American might have influenced participants classifying me as nonblack—a consequence of social whitening in Brazil—I do not believe that was the case. For, while most Brazilians did not see me as black, they generally did not see me as white. During my time in Brazil, people often assumed I was a poor, nonwhite Brazilian and treated me in a less respectful manner or questioned my presence when I entered (white) middle-/upper-class social settings, that is, until I disclosed my American nationality.[8] When I dined with white Brazilian and/or US friends in nicer restaurants, I was generally the darkest person in the establishment, and other customers stared at me, wondering why I was in those establishments. However, once I spoke English, emphasized my foreign-accented Portuguese, or mentioned I was in Brazil doing a Fulbright fellowship for my dissertation research at an elite US university, people quickly asked where I was from and were very friendly. This also happened when I socialized with white American friends and Brazilians who interacted with us initially thinking I was my American friends' maid.

I found that this differential treatment also extended into my fieldwork. I sometimes felt that Valadarenses were more willing to assist me in my research efforts and give me preferential treatment after I revealed that I was an American conducting research. At times, I got the sense that people wanted to befriend or become acquainted with me so they could say they had an American friend. In GV especially, all things that are related to the United States are given higher social status: Americans, Brazilians who migrate and return from the United States with high socioeconomic status, being proficient in English, or name-brand American products.

Sociologist Kelly Chong (2008) also discusses how being a US citizen from a prestigious university resulted in more admiration from respondents in her field site in South Korea due to US global dominance. Harrington (2003) and Horowitz (1986) also suggest that respondents might favorably respond to researchers when the latter's social status might enhance the former's social standing through association. In a few rare instances, participants assumed that I could help them get a visa to return to the United States because I was on a US government fellowship. As a result, I found myself in a precarious situation trying to negotiate my multiple and intersecting self-ascribed identities with the presumptions Brazilians made about me (and my social networks) based on my physical appearance and nationality.

After all, though I was a black American, I was perceived as a nonwhite and nonblack Brazilian. While I was sometimes mistreated because of my appearance, I was also a US citizen who could disclose that nationality and return to the United States whenever I chose, something that Brazilians of any color cannot do. I had the privilege of using my US nationality to gain social acceptance or better treatment when I felt people may have been discriminating against me unlike the brown and black Brazilians I encountered. Therefore, my American identity was a hidden one until I revealed it to people. Despite arguments that money socially whitens educated and middle-class nonwhites in Brazil and that it allows them to easily access privileged social treatment and white social spaces, I found that I was not "socially whitened" until I publicized my middle-class, highly educated status and American nationality. Thus, being perceived as a nonwhite "insider" allowed me to blend into the local population and get a better sense of how phenotype and socioeconomic status influence social interactions. My personal experiences and sociological interpretations of them were crucially important for understanding microlevel and interpersonal race relations in Brazil, which were important for analyzing the data I collected.

US Skin Color Stratification and Becoming a Brazilian Racial "Insider"

My experiences conducting research in Brazil helped me draw parallels between racial classification norms there and skin color stratification in the African American community. Like Brazilians, black Americans have a diversity of skin tones and hair textures given the race mixing (often nonconsensual) that occurred during US slavery. Thus, as a lighter-skinned black American reared in the southern United States, I was personally aware of this divisive and controversial issue, which has also been documented in various sociological studies (Wade 1996; Hill 2000; Herring et al. 2004; Hunter 2005). Because of my awareness of this issue, I noticed and observed the more explicit focus on skin color rather than rigidly defined racial groups in Brazil. In informal conversations, popular songs, telenovelas, and my interviews, Brazilians talked very openly and in much detail about people's skin tones, hair textures, noses, and lips, and it was clear that lighter or whiter phenotypes symbolized the ideal of beauty.

Therefore, Brazilians' discussions of skin tone and phenotype resurrected some very uncomfortable feelings and difficult childhood and adolescent memories of being treated favorably and unfavorably because of my lighter skin. The blatant privileging of lighter and whiter phenotypes in Brazilian media and society at times also made me conscious of how difficult it must have been to be a darker-skinned person in Brazil. I was again placed in a more privileged position than darker Brazilians (just as I sometimes still am compared to darker black Americans), as my lighter skin tone led others to not see me as black in Brazil. Cognizant of that privilege, I had internalized that most Brazilians did not see me as black although that was how I continued to self-classify. I realized that I had also internalized the relevance of skin tone among Brazilians and began to pay more attention to the variations in phenotype between individuals. Instead of seeing Brazilians through my "American" socialized racial lens, I had begun to view them with a Brazilian eye, and become a true "insider" in that regard.

Once I completed the research and returned home, during my readjustment period, I continued to pay attention to skin tones, not just of black Americans but of Americans of different racial groups, in a way that I had not done before. At times, I found myself wondering in which color (rather than racial) category Americans would be placed if they lived in Brazil, especially recognizing that black Americans with lighter skin would not be considered black in that context. Recognizing and coming to terms with this was difficult because I thought I had moved past the issue of giving relevance to skin tone as a black American woman in the United States. Living and conducting research in Brazil certainly helped me develop a nuanced understanding of the relationship between race and skin tone in Brazil and the United States. However, this experience also helped me recognize that the field site can transform how the researcher ascribes meaning to social constructions and processes, in this case racial conceptions. Although I returned to the United States with my black racial identity intact, having to renegotiate perceptions of my race, skin tone, and nationality in Brazil gave me a more nuanced lens for observing and researching race in the United States.

The Research Implications of Engaging with One's Positionalities

Negotiating multiple social identities can be a complicated process depending on the circumstance or context, especially when conducting qualitative research. Because social identities may become more pronounced in some instances and less apparent in others, social cues around us influence how those identities are activated (Stets and Burke 2000). While living and conducting fieldwork in GV, I felt conscious of all my identities all the time. I constantly thought about my racial identity as a black American and differences between race in the United States and Brazil. Whenever I passed a homeless person on the street, I became aware of my social class, reflecting on how I had the financial resources to live comfortably in another country. When I heard stories of deportation and the struggles of obtaining a tourist visa as a means to immigrate to the United States, I was conscious of being a US citizen, which allows me to travel internationally without drawing suspicion that I will become an undocumented immigrant. When having discussions with Brazilian women about the blatant sexism and machismo in Brazilian society, I was aware of my gender. As a speaker of three languages interviewing Brazilians who only spoke Portuguese, I was reminded of the educational opportunities I had which allowed me to learn those languages. As I learned more about Valadarenses through our conversations and interactions, I reflected on my various social identities and how they influenced my perception of the world.

Likewise, my positionalities challenged the preconceived notions that Valadarenses I interviewed had about me, as a middle-class "Brazilian-looking" American. Our exchanges shifted how I engaged with my research because they provided relevant insights into Brazilian society. Ultimately, this allowed me to better understand the people and the city where I was conducting fieldwork.

For researchers, particularly those who do qualitative research internationally or domestically, it is imperative that we make an earnest effort to truly understand as much as possible about the social context where we do our research. While background reading and extensive study are a useful introduction, they are not sufficient to comprehend the numerous social processes that occur in the field where we derive social inquiry. It is essential to spend some time in that setting and, if possible, to live there, interact with the people, and get involved in that social setting. Getting involved necessitates sharing one's self with the research community, which is hindered if the researcher does not speak the language of local residents. Through living in GV, I could see how my own social identities and perceptions of the world differed from those of the people I interviewed. In recognizing those differences, I was able to develop a better understanding of how race functions in Brazilian society, which was necessary for my project.

As researchers, much can be gained from carefully and personally putting our "social" selves into the research process. After all, we are social beings and our humanness is what constitutes our ability to ask the questions we want to learn about in our research. However, this does not mean we should allow our identities and personal perceptions to dictate our research findings. While conducting the research, I constantly wrote field notes, as most qualitative researchers do, to describe everything that was happening around me during interviews and also during informal social interactions. When analyzing the interviews and reporting the findings, I often referred to those field notes to minimize my US-based and sociologically trained interpretation of the data. Doing so transported me back to the field and helped me ground the research in the racial context of GV and Brazil. I had also posed open-ended and closed-ended questions to respondents, which allowed me to assess consistency, patterns of evidence, and counterevidence in their answers. Furthermore, continually reviewing interview protocols and field notes and listening to the interviews helped me keep the project grounded firmly in respondents' experiences.

My physical, social, and cultural immersion into the field site was crucial for the interpretation of my data and my development of a theoretical framework to explain how the experience of immigration and return migration transformed my respondents' and my own racial conceptions. I refer to that framework as the "transnational racial optic," which I define as a "lens through which migrants observe, negotiate, and interpret race by drawing simultaneously on transnationally formed racial conceptions from the host and home societies" (Joseph 2015, 7). My conversations with return migrants revealed that they transported Brazilian racial conceptions in terms of flexible racial classifications and cordial social interactions with them to navigate US race relations, which they found to be more exclusive and overtly discriminatory. However, during their time in the United States, they adapted and acquired US racial norms, which then transformed how they viewed Brazilian race relations after returning. Despite migrants' difficulty in navigating US categories and experiencing discrimination, many observed a sizable black middle class, which they felt indicated that the United States had more social mobility *across* racial groups, whereas in Brazil, black (and brown) Brazilians experience much more social, political, and economic marginalization relative to white Brazilians and black Americans. Thus, migration produced a hybridized Brazilian-American lens or transnational racial optic for recalibrating these individuals' understanding of race in each country. My return migration also allowed me to recognize differences in the structural positions of racial groups and manifestations of racism in the United States and Brazil. Like my respondents, living abroad had changed me and I too experienced the transformative impact of the transnational racial optic.

Conclusion

We researchers are social actors who cannot completely disentangle ourselves from the social world in which we do research. However, this does not mean that we allow our positionalities to guide our findings. We must be aware of them and how they influence participants' engagement with us in the field. It is also important to allow participants' perceptions and experiences to guide the research process. Conducting research on the racial conceptions of Brazilian return migrants in GV brought unique personal and scholarly challenges because I was doing fieldwork in a context that was socially, racially, and culturally different from the United States. My identities as a native, middle-class, black American woman influenced how I socially navigated GV and made sense of my surroundings, and how people made sense of me. In order to complete the project, I had to become familiar with my research setting to understand its social and cultural context and recognize how my identities and perceptions differed from those of the people in GV. Doing this helped me develop the appropriate social lens for interpreting my data so I could analyze and report my findings as accurately as possible. My engagement in the field also aided my development of social theory to assess how movement across borders transforms how individuals see and understand race. A similar process of engaging with one's positionality and fieldwork context is essential for qualitative scholars conducting research domestically or internationally.

Notes

1. I also interviewed a comparison sample of 24 nonmigrants, that is, individuals who never immigrated, to more effectively assess the influence of migration on returnees' racial conceptions. For more information about and empirical findings from the project, see Joseph (2015).

2. I use "American" throughout the chapter as an adjective to describe my positionality as a US citizen for simplicity purposes. Participants in the research site also referred to me as an "American" even when I self-identified as "Estado-unidënse," the Portuguese word for people from the United States.

3. Some famous personalities who have a skin tone similar to mine are Halle Berry and President Barack Obama. Although these two individuals are biracial, they are perceived as black in the United States and likely would not be perceived as black in Brazil.

4. Even foreigners (e.g., other Americans, Europeans) vacationing in Brazil often assumed I was Brazilian, especially when I spoke Portuguese with Brazilians.

5. When I called respondents to schedule interviews, I explained that I was an American graduate student conducting research with Brazilian return migrants. Therefore, it was clear that my phone identification as American yielded a presumption that I was white, which contrasted with my physical appearance when I showed up for interviews.

6. At the end of the interviews, I asked all 73 respondents how they racially classified me by Brazilian Census and open-ended categories. Using the former, 37 percent classified me as pardo (brown), 28 percent classified me as black, and 31 percent classified me as "other." Two respondents classified me as white. The most common open-ended categories I was classified as were morena—mixed/brown, "café com leite"—coffee with milk, and "jamba"—the color of a rare reddish brown Brazilian fruit.

7. Each quote is translated from Brazilian Portuguese and includes the participant's self-ascribed racial classification, gender, and age when I interviewed them.

8. This was especially the case if I was casually dressed.

CHAPTER 6

Guess Who's Coming to Research? Reflections on Race, Class, Gender, and Power in Salvador, Bahia, Brazil

Jaira J. Harrington

I am often asked about how I developed an interest in paid domestic work,[1] as I wrote both my master's thesis and dissertation on the subject.[2] Quite honestly, it was difficult not to be interested as a black woman researcher in Brazil. According to the most recent reports from the 2010 PNAD-IBGE, Brazil's largest national household survey, over 7 million individuals, about 93 percent of which are women, are employed in paid domestic work. A 2011 study conducted by the Departmento Intersindical de Estatistica e Estudos Socioeconomicos (DIESSE) shows that nearly one-fifth (17 percent) of employed black women work in paid domestic positions in the largest metropolitan areas of Brazil. In terms of the racial makeup of black domestics in comparison to nonblacks, the distribution of women ranges from 97 percent in Salvador to 49 percent in São Paulo. Black women are significant, if not overwhelming, contributors to this labor market in all of Brazil's largest municipalities.[3]

In my early Portuguese language study and research on black women's organizing within the Unified Black Movement, I lived in city center neighborhoods including Aclimação and Bela Vista, in São Paulo; Flamengo in Rio de Janeiro; and Barra in Salvador. As a black woman living in the city center, residents, tenants, business owners, and other fixtures in the neighborhood considered me a transient visitor and not a resident. As a black woman, my natural domain would be in the social and geographic periphery. Presumably, the only logical reason that I would be in that space is to perform work as a service laborer, principally as a domestic worker. I know through conversations I had with my neighbors and doormen after uncomfortable encounters regarding the social and service elevators that the first one was reserved for residents and guests and the second for the laborers in the building.[4]

Open, limited, and restricted access to spaces was critically important and instructive on various levels. On the one hand, as a guest researcher, I had the privilege of a US passport and access to institutions of higher education with my University of Chicago pedigree; on the other hand, I still negotiated ever-present discrimination and social dislocation as a black woman living in Brazil. Additionally, within this spectrum there were daily experiences that did not fit neatly into this binary description of accommodating both situations at the same time.

As other narrations in this edited volume reveal, the theme of "rightful place" is a fairly common and notable feature of transatlantic research. I often feel that, as a fellow person of African descent and a researcher, this dislocated feeling parallels the lives of the people who bring my research to life. For example, Ana Flávia Peçanha de Azeredo is a woman that orients Hanchard's (1994) "Black Cinderella? Race and the Public Sphere in Brazil." While visiting a friend in a residential building of Vitória, Espírito Santo, Peçanha was viewed as a black woman out of her rightful place because she took the social-class elevator instead of the service-class elevator and was verbally and physically assaulted by the residents. The story received national attention because she was the daughter of the then governor, Albuino Azeredo. Hanchard links her story to the broader myth around racial democracy and the Afro-Brazilian experience of marginalization and discrimination. Some 20 years later, the account of a woman called "black Cinderella" by the news media still resonates with the interrelationships of race, labor, gender, and power.

There is recent scholarly work on justifications for racialized and gendered violence. Smith's article (2014) on the 32-year-old Rio de Janeiro domestic worker reflects upon the case of Sirlei Dias de Carvalho Pinto.[5] While waiting on the public bus in the early morning hours in Rio de Janeiro, Pinto was assaulted, robbed, and severely beaten by a group of young adult men. The media reports suggested that the severity and brutality of this crime was shocking given the middle-class status of the five men involved. The men justified their behavior to the police by alleging that they thought that she was a prostitute. In the days following the assault, multiple parties commented on the nature of violence against women. Pinto herself stated: "I forgive them as fellow human beings, but it is still important that they pay for what they have done to me."[6] Despite her horrifying experience, which warrants legal intervention, she still maintains an empathetic, humane view of her assaulters.

The story of black women out of their rightful place is a familiar concept to black bodies within the African diaspora, especially in Brazil (Bairros 1991a, 1991b; Carneiro 1999; Caldwell 2004, 2007; Perry 2004, 2008, 2013; McCallum 2007; Henery 2011). Yet the distinct relationship of the researcher's racialized and gendered body to fields of knowledge is still an understudied issue. The central aim of this chapter is to elucidate the implications of power and aesthetics on the field of qualitative field research from the lens of the researcher, in this case, myself. I elaborate these issues using three accounts—(1) homestay living, (2) language school training, and (3) qualitative field studies—from my experiences living abroad as these themes relate to my current research agenda around domestic work.

The title of this chapter "Guess Who's Coming to Research?" is a bit tongue-in-cheek in that I am an invited guest to the country, an American researcher, but as I demonstrate, the black, female body that arrived to conduct research was not the expected, traditional body to do so. Similar to the stories of Peçanha, Carvalho, and countless other black women, this hesitant, tentative welcome was the fountain from which my research interests in the marginal labor category of domestic workers flow. This chapter is organized in the following manner: first, I discuss Salvador as a research site; second, I provide brief auto-ethnographies of three cases; and I conclude with directions for future research.

The Salvador Case

In 2010 and 2012–2013, I had language training and conducted research on paid domestic work in Salvador, Bahia, Brazil. As the first colonial capital and port of entry for enslaved Africans in Brazil, Salvador is a popular research site for the trajectory of historical and contemporary race, politics, society, and economy narratives (Castro 1993; Borges 1993; Butler 1998; Kraay 1998; Perry 2004, 2008, 2013; McCallum 2007; Mitchell 2009; Pinho 2010; Romo 2010; Mitchell-Walthour 2011; Hordge-Freeman 2013; Williams 2013).

For my research on the race and gendered aspects of organized domestic work, Salvador presents an ideal paradox. Not only is the economy of unpaid and low-wage black labor a historical element of this site, the city is also the nucleus for the marginal category of domestic workers through the National Federation of Domestic Workers (FENATRAD). Led by the well-known activist Creuza Maria de Jesus, this organization represents and articulates the rights of domestic workers. So while black labor has been historically undervalued, this city has nurtured a national flagship that has helped to augment predominantly black and female domestics' occupational rights.

These contextual details of Salvador constitute how the gendered black body is understood within the city. These *imagined possibilities* or the manner in which these societal limitations and expectations were imposed on me gave a particular insight into my research: urban, labor, and race politics. While conducting research, my black, female body gave me direct contact with the presumption of rightful place and open, porous, or limited barriers to entry within the cityscape. The perceived social status of an individual embedded in the social context judged by their appearance is what I call *aesthetics of power*. As a black woman, I was frequently interpreted to have very little power. The sections that follow will elaborate upon interrelationships between these contextual elements of imagined possibility and aesthetics of power with three examples directly from the field.

Welcome to Salvador! The Host Family Experience

While living with a host family in Salvador, I had daily encounters with imagined possibilities and power aesthetics. From the very first day I arrived at my host family's residence, I learned how I was perceived in the wealthy, touristy Barra neighborhood. Excited about my first stay in Salvador and advanced language courses, I had previously reached out to my host family through e-mail and had some fun conversations with them. As the taxi pulled up to the apartment complex, my host mother exited the gate. I hopped out of the taxi to meet the family matriarch, as I felt that I had already known her for some time. So when she had a distant and confused look while asking the taxi driver if I was "the American student" in Portuguese, I wondered who else I could be on the expected date of arrival ringing the doorbell. I struggled to divorce myself from the idea that perhaps I was a misfit to her idea of what an exchange student should look like. She might have thought that people of African descent are not doctoral students from US universities with whom one exchanges friendly e-mails. Somehow, my physical being ran contrary to her imagined expectations. It was an uncomfortable moment. Like other experiences in Brazil, should I paper it with a laugh or take note of the histories of the past as they unfold and shape the present?

As a black woman studying in elite language schools and living with wealthy host families that are predisposed to a particular global North, white-normative, and/or male aesthetics emblematic of wealth and power, I had often been confused for a housekeeper or domestic worker while occupying these spaces by neighbors and domestics working in the households in which I lived. Throughout my summer there, my host mother regularly confused my name with that of her domestic worker, which struck me as strange. Although we were both unambiguously of African descent, our physical characteristics were not at all similar. Our names could not be more dissimilar. This almost daily moment of discomfort was exacerbated by my first memory of my host mother at the taxi. Is it possible that she wanted to remind me of my rightful place, which would be a position of service in her household? Could she only understand limited possibilities for my life and existence as a black female body? Could it have been a simple slip of her speech? Whatever the case might be, these experiences became an extension of my research.

Class Is in Session: Race, Gender, Place, and Power

That same summer of 2010, I began advanced Portuguese language courses. "Tudo bem? Qual é seu nome?" a French student with a heavy accent cheerfully asked my name. I told him my name and background. I am from the United States of America, doctoral student of Political Science and researcher. His eyes lit up in surprise: "Oh, I thought you worked here!" I was rather excited to hear him say that. To me it meant that I might have improved my Portuguese to such an extent that my accent was slowly dissolving. Maybe I used the subjunctive correctly and the proper masculine and feminine articles for nouns. It was an exciting moment!

Out of curiosity I asked: "Did you think I was a professor?" He replied: "No." So, I pressed further: "So, an administrator?" He said: "No. I thought you worked here." Then, I realized what he meant. He thought that I might have been part of the housekeeping staff. I confidently replied: "No, actually I'm studying here, just like you."

Having only arrived a few days earlier, this French student already understood how Brazilian aesthetics of power operate. A black woman in this setting could not possibly be a student, potential client, teacher, or administrator. The only plausible possibility for my life was in a position of service.

This exchange with a fellow student is one experience that often comes to my mind, although it certainly was not an isolated experience. Not only do aesthetics of domination exist in Brazilian life as it relates to domesticity, but also aesthetics of power in daily life.[7] In my case, these issues were intertwined. However, I was not disheartened by this experience, as it only informed my broader research agenda on domestic work. Having some basic understanding of the perceived and real limitations of black women's lives through these restrictions imposed upon me revealed streams of knowledge that I would not otherwise have access. My interview work within the union clarifies this point.

Research in Salvador Domestic Workers' Union

I conducted qualitative interviews at the Salvador Domestic Workers' Union for dissertation research in 2012 and 2013. In my previous experience in Salvador, I lived in Barra, a wealthy neighborhood with a white family. This time also I rented a room from a working-class white family in the working-class neighborhood of Fazenda Garcia. I regularly walked to the union to conduct my interviews. In this neighborhood, I was not confused for a non-resident and did not feel uncomfortable or out of my place.

While speaking with the leaders of the union and participating in their socials, presentations, and other activities, the women would commonly frame their conversations with "we" so that I would identify with some of the issues with which they contend as black women. In interviews with domestic workers, sometimes the women would also employ "we" within their response to express their understanding of a shared experience of black womanhood. While living in neighborhoods as a black woman created distance and isolation, that same identity opened up doors of opportunity. As a researcher, I gained a sense of empowerment among this collective. I could only imagine the impact of organizing on the group participants.

Conclusion

The purpose of this chapter is to elaborate upon my experiences as a researcher in Salvador, Bahia, Brazil. I show how my body incurred racialized and gendered inclusion, exclusion, and placement within the city of Salvador. This dimension of my research provided me with an opportunity to access new fields of knowledge in terms of imagined possibilities for my existence and the presumed power implied by my aesthetics as a black woman. In the homestay and language school, the possibilities for my life were limited, while in the qualitative research with domestics I was invited into their world through a parallel exclusion.

Yet, I am not naive. There were privileges to my US nationality and temporary researcher status that made my female blackness more acceptable to others. I could ostensibly pack my bags and go at any time to deal with a more familiar racism at home. I could notify people that I was not Brazilian and that I was indeed American and lavish in the privilege of being a "special" kind of gringa—a term for foreigners that I regularly rejected on the foundation that I was aware of my privilege and I exercised the right to deploy it strategically. The women with whom I spoke at the union did not have that option. In my own research, I saw power differently within the women with whom I worked and interviewed.

To some degree, this experience was not unique to Salvador, as I have researched in other cities in Brazil with similar experiences. Although I focus on field research and scholarly activities, further research in this area might extend to traveling while black within Brazil. An approach using the analytic lens of imagined possibilities and power aesthetics could help us to understand black leisure culture as well.

Another future direction for research could touch upon the emotional consequences of placement. The instability of these categories presents emotional costs beyond homesickness that are ripe for extended discussion and research. Emotional resilience of the researcher is a strong factor in conducting and completing this kind of research in these conditions. In sum, there is a clear place for specialization on race, gender, class, and power, yet the social context and complexity related to these issues must be taken into account.

Notes

1. To honor the skills and the work of domestics, I embrace the term "domestic work" instead of "domestic service." The use of the word "service" may impute societal and gendered expectations of work upon the women who perform labor. Domestic workers are particularly vulnerable to the unclear and ambiguous nature of their work, which has been viewed as both a labor of care and inherently a women's work that is often underpaid, undervalued, and unrecognized. The word "work" more distinctly valorizes these women's profession. Domestics are women who either live in the home permanently or visit their employer's home frequently to carry out household chores for families and individuals. In addition to housework, the women often have a hand in raising children and caring for the elderly. Although domestic work in Brazil may involve men, this situation is highly unlikely. It is more often the case that women are paid household domestic workers. Various sources (DIESSE 2011; Hite and Viterna 2005) report that women occupy up to 95 percent of domestic work positions in Latin America and Brazil.

2. See Harrington (2010, 2015).

3. See Barbosa et al. (2003); Pinho and Silva (2010); Rezende and Lima (2004).

4. This tradition of architecture also seems to undermine the notion that formal segregation did not exist in Brazil and it was only based on class. If that were the case, I had the privilege of US nationality and the status of a visiting international scholar, but these attributes were not salient and immediately perceptible. Therefore, a strong reason for being directed to the service elevator would be because of my black, female aesthetic.

5. "OAB-RJ compara violência com empregada doméstica ao caso João Hélio." *UOL Última Instância*, June 25, 2008, Online Edition.

6. Freire (2007).

7. For an extended discussion on aesthetics and domestic work, see Goldstein (2003).

CHAPTER 7

But You (Don't) Look Like an African American: African Diaspora Looking Relations between Brazil and the United States

Reighan Gillam

In November 2007 in São Paulo, Brazil, I attended a lecture by Fred Hampton Jr., the son of slain Black Panther Fred Hampton Sr. Hampton's lecture was one of the many events about black culture, history, and politics that took place in São Paulo during November, or the month of black consciousness. I attended the lecture with Manoel and Christina, two Afro-Brazilians around my age who worked and studied in São Paulo. Christina's mother also attended the event, but she arrived at the venue, a local community center, before we did. She had saved seats for us in the crowded space to ensure that we could all sit down to hear the lecture. I sat between Christina and her mother. After engaging in a series of small talk, Christina told her mother that I was from the United States and that I had come to Brazil to examine racial politics. On receiving this information, Christina's mother touched my arm, pointing to my skin color, and said: "I wouldn't think you were from the United States, you aren't that dark (*voce não é tão pretinha*)." In referring to the relative lightness of my skin, Christina's mother reflected the common belief in Brazil that Brazilians are of mixed race and African Americans,[1] or US Americans in general, are racially singular. To her, the idea that African Americans are racially pure or singular would manifest itself in a phenotype of dark skin and coarse hair. Yet I, as an African American with light brown skin and curly hair seemed to fall in line more closely with her ideas of how Brazilians looked.

How Afro-Brazilians look at African Americans, through lenses ideologically structured by popular theories of race and nation, is the subject of this chapter. Brazil's national emphasis on racial mixture was established and gained hegemonic power partially through its comparison to the United States as racially binary (Seigel 2005). But how do these different national racial ideologies inform popular aesthetic understandings and mediate encounters between people of African descent from the United States and Brazil? Drawing from 18 months of fieldwork between 2007 and 2013, my chapter will ethnographically explore how ideas of Brazilian mixture and US American racial singularity provide the physical coordinates upon which understandings of each group are visually mapped upon the other. During my interactions with Brazilians of African descent at black political events in São Paulo, Brazil, and during everyday encounters I noted and tracked the constant references to racial ideologies in the United States and how I (or others) did or did not deviate from the supposed phenotypic categories of blackness thought to accompany each country. Some Brazilians of African descent interpret the US racial ideology of binary racial categories to mean that African Americans are not racially mixed and are therefore dark-skinned, with coarse hair, and with what are perceived to be stereotypically black physical features. Encounters between African Americans and Afro-Brazilians sometimes served to disrupt this idea by inferring that people of African descent in Brazil and the United States do not look that different. However, I contend that these interruptions remained partial due to their seeming inability to produce an understanding that Brazil is not exceptional in having a racially mixed populace since miscegenation or racial mixture is a pattern found in slave societies throughout the Americas. Additionally, during my interactions with Afro-Brazilians in São Paulo, dark skin and black physical features remained spatially fixed in the United States or located in Salvador, Bahia, Brazil, leaving them as exceptional looks for a nation of mixed races. I use the term "African diaspora looking relations" to describe the evaluations made between people of African descent who are informed by assumptions and ideas about phenotypic differences thought to correspond with ideologies of race and nation. I use the case of African American and Afro-Brazilian interactions, rendered through ethnographic detail, to understand how processes of identification can disrupt or maintain national ideologies around race. At stake is the disruptive potential that black encounters can have for racial ideologies.

Brazil has commonly been characterized as a racial democracy, where centuries of racial mixture have resulted in a color spectrum that blurs the boundaries between racial groups. This absence of discrete racial categories, such as black and white, is also thought to prevent racism and racial inequality in Brazilian society. Gilberto Freyre is credited with popularizing and propagating this ideology. His book, *The Masters and the Slaves* (1944), located the origins of Brazil's mixed-race population in sexual relations between enslaved African women and their Portuguese masters. He also argued that this racial intermingling made possible Brazil's cordial race relations and diminished the presence of racism in society. This ideal of harmonious race relations resulting from centuries of racial mixture continues to structure many Brazilians' contemporary attitudes to race relations and racial dynamics.

The idea that Brazil is populated with racially mixed individuals translates aesthetically into the overvaluation of brownness or *morenidade* in Portuguese. Patricia Pinho points to the "ubiquitous images of *gente morena* [brown people] as quintessential Brazilians" (2009, 48). Pinho uses the image of Gabriela, the female protagonist of the popular Brazilian book *Gabriela, Clove and Cinnamon* (1962) (*Gabriela, Cravo e Canela*) by Jorge Amado, to demonstrate how "these gendered representations of mestiço [mixed race] phenotype already circulated in the commonsense imagination" (48). Pinho writes:

> The cinnamon-colored skin of Amado's title character exhales the smell of cloves. Flavored by these spices brought to Brazil by the Portuguese colonizers, Gabriela literally embodies the national myth of Brazil's peaceful and racial coexistence and amalgamation, enacting Freyre's notion that mestiço types are "on the aesthetic level, plastic mediators between extremes" (Freyre 111). (48)

Brownness, or an aesthetically ideal mixture of black and white, mediates Brazilian conceptions of who looks like a "typical Brazilian." Brownness continues to be promoted as a physical ideal through actresses like Sonia Braga and Camila Pitanga who "embody brown standards of Brazilianness" (Pinho 2009, 47).

Although Brazilians consider themselves to be a nation of mixed-race citizens, which prevents color lines from being drawn, scholars have found that inequality can be traced along racial lines (e.g., Silva 1985; Paixão 2004; Telles 2004). These unequal racial lines are expressed aesthetically through the devaluation of black features, such as coarse hair (Burdick 1998; Caldwell 2004). Pinho narrates the flipside of Brazil's infatuation with brownness or its physical ideal of mixture: "[W]hile antiblack racism in Brazil is expressed mainly against dark-skinned individuals, it also operates in the devaluation of physical traits 'deemed black' even in those who have lighter skin complexion, thus creating 'degrees of whiteness'" (2009, 40). These traits that are "deemed black" include wide noses, coarse hair, full lips, and brown or dark skin color. Elizabeth Hordge-Freeman extends Pinho's analysis by examining how Afro-Brazilian families evaluate racial features in ways that "can reproduce a white supremacist ideology by rewarding proximity to whiteness" (2013, 4). Hordge-Freeman found that the Afro-Brazilian families she worked with in

Salvador, made aesthetic judgments that, oftentimes, devalued the coarse hair or wide noses of their children and family members, thus showing that "racial socialization in families can naturalize whiteness and promote the devaluation of black looking features" (12). Hordge-Freeman did find that families are complex and also promoted racial pride. Together, Pinho and Hordge-Freeman present the aesthetic coordinates in Brazil through which whiteness maintains a hegemonic standard of beauty. Yet, the proximity to whiteness can be achieved through brownness, or some combination of brown or light skin and, more importantly, features that are less marked as black. This aesthetic ideal of brownness also informs a hegemonic image of the archetypal Brazilian. But not everyone, in fact few people, can achieve this ideal of brownness resulting in the stigmatization of their "black features."

Ideals of Brazilian "mixedness" or brownness are also maintained through the comparison between Brazil and the United States. Micol Seigel demonstrates that the common juxtaposition between racial systems in Brazil and the United States has left the impression of "the United States as a place of overt racism and a stark, dichotomous racial system and Brazil as a place of subtle, gradated multiplicity" (2005, 67). The United States then is not only the place of racially binary populations, but also the place where racism is thought to be stronger. Robin Sheriff writes that "televised images of racial violence occurring outside of Brazil buttress the notion that racism is a particular kind of social and moral problem to which Brazil has always been, and remains, relatively immune" (2000, 120). Images of the United States as racially binary and Brazil as racially mixed go hand in hand with ideas about cordial race relations in Brazil and more hostile race relations in the United States. Evidenced by the opening anecdote of this chapter and with the ethnographic material presented later, I contend that the idea of the United States as racially dichotomous and Brazil as racially mixed carries the phenotypic assumptions that African Americans are darker than Brazilians and have facial features that are associated with blackness, while Brazilians of African descent are generally lighter-skinned or adhere more closely to a standard of brownness. These ideas of Afro-Brazilian brownness and African American blackness belie the reality on the ground that racial miscegenation during and after slavery took place in both countries and that African Americans and Afro-Brazilians display a stunning range of phenotypes, including varying combinations of skin color, eye color, hair texture, nose shape, and body size. This chapter explores how Afro-Brazilians and African Americans look at one another through lenses ideologically structured around ideas of Brazilian brownness and African American blackness. Before turning to Afro-Brazilian and African American encounters, I ground their interactions in more general processes and histories of African diaspora formation.

Brazil and the United States in African Diasporic Perspective

African American and Afro-Brazilian interactions can be considered part of the processes through which African diasporic relations are made. The African diaspora refers to the scattering of African-descended people from Africa to the rest of the world through processes of slavery and migration. Paul Gilroy's *The Black Atlantic* (1993) expanded upon the idea of the African diaspora to include the processes through which black people have endeavored to create shared imaginaries of belonging based upon race and blackness across national boundaries and other geographic limitations. He takes his figure of the slave ship and traces its route through the Atlantic to discuss the global circulation of black people, cultural products and things, and the kinds of possibilities these movements engendered. Scholars have theorized diaspora as a relation (Brown 2005), citizenship (Boyce Davies and M'Bow 2007), and an encounter (Caldwell 2007).

Building on Gilroy's work, Edmund Gordon and Mark Anderson call for ethnographies of what they term "African diaspora identification" (1999). Gordon and Anderson define "African diaspora identification" as "not so much [a focus] on essential features common to various peoples of African descent as on the various processes through which communication and individuals identify with one another, highlighting the central importance of race—racial constructions, racial oppressions, racial identification—and culture in the making and remaking of diaspora" (284). Gordon and Anderson contend that we shift away from the attention previously given to finding cultural patterns of commonalities and differences between black groups or the same cultural patterns, beliefs, and activities between African-descended groups and those in Africa, often called Africanisms. Instead, they argue that scholars should focus on how different black groups identify with each other.

As anthropologists, Gordon and Anderson call for "ethnographic attention to the process of African diasporic identification" (284). They write that "rather than assigning identity and positioning how people should participate in the making and remaking of diaspora, we must investigate how they actually do so" (284). Ethnography, and its commitment to describe experiences, actions, and talk through the observer's lens, presents the means through which to render the lived relations of African diasporic identification. Rather than presuppose solidarity, animosity, or indifference between black populations and groups, Gordon and Anderson call upon engaged observers to depict the lived experiences and everyday unfolding of actual encounters between African-descended populations. Gordon and Anderson provide a useful concept to understand black encounters, yet they do not account for the ways in which African diaspora encounters can be structured by assumptions embedded within ideals of race and nation.

Before my examination of the African American and Afro-Brazilian encounters I observed, it is important to note that relations between these groups are not new and occur through various means. They have glimpsed one another's struggles and cultures through the black press in São Paulo (Seigel 2009), roots tourism (Pinho 2008), Soul music (Fontaine 1981; Scott 1998; McCann 2002; Alberto 2009), activism (Vargas 2003), gospel music (Burdick 2013), hip-hop music (Perry 2008; Reiter and Mitchell 2008), and Obama's presidential victory (Gillam 2013). David Hellwig's *African American Reflections on Brazil's Racial Paradise* (1992) uncovered a historical pattern of African American engagement with Brazilian ideas of racial harmony. From about 1900 to 1940 African Americans affirmed Brazil's idea of racial democracy, from 1940 to 1970 they questioned it, and from 1970 until the present they deny it. Hellwig compiled writings about Brazil by African American intellectuals and writers such as W. E. B. Du Bois, E. Franklin Frazier, Angela Gilliam, Lorenzo D. Turner, and Rachel Jackson Christmas who follow these patterns.

Film director Thomas Allen Harris offers a cinematic rendering of the encounters between African Americans and Afro-Brazilians in his film *That's My Face* (2001). In the film, Harris chronicles his spiritual journey to Brazil, where he went in search of the Orishas or the deities of the Candomblé religion, an African-descended religion in Brazil. Harris grew up in a protestant African American family in the Bronx, New York, where he was exposed to global ideas of blackness. The film reveals that his grandfather always wanted to travel to Africa and he would constantly expose the family to films and television programs about Africa. Harris had the opportunity to realize his grandfather's dream when he went with his mother and brother to Dar Es Salaam, Tanzania. They lived there for two years while his mother taught at a school. When they returned to the United States, Harris's mother married a South African man exiled in the United States due to his anti-Apartheid activities. A Kenyan friend of Harris introduced him to Conceição, an Afro-Brazilian woman who educated Harris about the Orishas of Candomblé. This meeting prompted Harris to travel to Salvador to search for his ancestral spirits and become more familiar with the Candomblé religion.

Sheila Walker describes Candomblé as "the religious system evolved by enslaved Africans and their Afro-Brazilian descendants" (1990, 103). Walker defines the Orishas as

the anthropomorphized forces of nature that are the spiritual beings of the Candomblé, and the intermediaries between the creator and his human creations. The Orishas came to Brazil during the transatlantic slave trade with the Yoruba people from present day Nigeria and Benin, the African ethnic group whose religious culture has remained most intact and influential in both Brazil and elsewhere in the Americas. (104)

In the film Harris states: "I see myself in the Orishas, in their duality." He is shown attending a public ceremony of ritual offerings to the Orisha Iemanja at a beach in Salvador, going to a location where animals are sold for Candomblé rituals, and eventually visiting a Candomblé religious leader called a babálorixa who does a reading for him and reveals his Orisha to be Oxun. Harris's journey to Brazil depicts an African American man's encounters with Afro-Brazilian people and culture through the lens of spirituality and religion. Harris identifies with Candomblé as a religion descending from West Africa that is continually created in Brazil. Harris's trip to Brazil presents one node in a series of interactions between African Americans and Afro-Brazilians. *African American Reflections of Brazil's Racial Paradise* (1992) by Hellwig and *That's My Face* (2001) by Harris privilege African Americans as the source of the transnational black gaze. In the next sections I turn to ethnographic examples of how Afro-Brazilians view African Americans and the ideological implications of such views.

On Looking Like a Brazilian

In June 2013, I attended a book launch party at the Pontifical Catholic University (PUC) in São Paulo, Brazil. During the event, the book's author and contributors summarized the book's content and chapters, which were about black culture in Brazil. A film about a remote black community in São Paulo's interior and a musical selection accompanied the book discussion. Afterward, as is customary at the many book launches I have attended, wine was served and attendees mingled in the lobby where the book was being sold.

This was the first event I had attended during this particular research trip and, as I had just arrived in São Paulo, I had not yet contacted any of my friends. So I was pleasantly surprised to run into Mariana, a friend since 2007 in São Paulo. Mariana self-identifies as Afro-Brazilian and has been involved in black politics in the city. For example, she had been teaching classes through law 10.639, which requires Afro-Brazilian history and culture to be taught in all public schools. She instructed teachers on the information about Afro-Brazilian history and culture that should be included in the classroom and techniques for teaching it. As the guests mingled Mariana and I caught up from when we had last seen each other. As she introduced me to other people, she revealed to them that I was from the United States. "Wow, I would never guess that you're from the US," said one of Mariana's friends. "Well we don't look too different in the US than you do here," I responded.

Mariana said: "No you do look more Brazilian. I don't know what it is . . .
maybe it's your way of being (*jeito de ser*) . . . I don't know." Then she followed
up saying: "You have the face of a Brazilian." I asked: "What does that mean?"
Mariana's friend answered: "It's your hair . . . because it's curly it makes you
look more Brazilian." An Afro-Brazilian professor from the university joined
the conversation and told us about the six months he spent in Texas conducting a research project on black music. "You know," he said, "about 25% of
African Americans are mixed with native Americans." "Oh!" Mariana
exclaimed. "Are you mixed with anything?" she asked me. I told her I was
mixed with white ancestry since my grandmother's father is white. "That's why
you look like that, that's why you look more Brazilian," said Mariana. "Most
African Americans are mixed with something," I responded. My response
echoes Anthony Marx's finding that "between 75 and 90 percent of African
Americans now are estimated to have some white ancestry" (1998, 70).
Mariana said: "Oh see we don't know this. People from the black movement in
Brazil talk about the US as if you all aren't mixed." The professor chimed in
saying: "I think it's really a matter of ideology. People basically look the same,
but it's the way blackness is defined in the two countries that's different."

During this exchange, Mariana and her friend thought that I looked
Brazilian because I embody a brown aesthetic because of my brown skin and
curly hair. The professor added some context to the conversation by confirming that African Americans are also racially mixed, but blackness is traditionally defined in the United States through hypodescent, or the "one
drop rule." Marx writes that "after the mid-nineteenth century the offspring of interracial unions were generally categorized as blacks, even if they
had 'one drop' of African blood" (69). This revelation called into question
Mariana and her friend's insistence that only Brazilians occupy a brown
phenotype. I tried to further unsettle their ideas by revealing that my grandmother was mixed. The professor and I tried to convey that the reality of
African American and Afro-Brazilian similar phenotypes contradicts the
ideological position of Brazil as the place of racial mixture and the United
States as the place of racial binaries.

The conversation shifted then and Mariana pointed to two women standing near the wall. They were dark brown–skinned with natural hair styled in
low Afros. "See, they look like they're from Salvador, Bahia," said Mariana.
She continued: "Yeah, they must actually be from there." Because they were
darker-skinned and had coarse hair Mariana located them within the geography of Brazilian blackness that places Salvador a city in the northeast of Brazil,
as the center and fount of black people and black culture. Before we could
continue down that line of thought, the two women came over and were
introduced to us by someone else. The introduction revealed that the women

also came from the United States. They had brought a group of students down from their college to study abroad for about three weeks in Brazil. Neither Mariana nor anyone in the group said anything about the revelation that they were not from Salvador, or even from Brazil, but from the United States.

In the above exchanges, phenotype and physical features mediated all of the interactions and assumptions about other people's origins. The revelations of African Americans who look like Brazilians seemed to momentarily interrupt ideas of "pure black" African Americans. Yet, no one appeared to reach the conclusion that perhaps African Americans and Afro-Brazilians do not look that different. Both countries have histories of racial mixture during and after slavery and African Americans vary in color and complexion. The professor tried to insert an ideological reading of the situation into the conversation but this did little to unsettle their logic of national difference in appearance between these two black populations.

Calling Brazilian Mixture into Question

In July 2013, São Paulo along with other cities in Brazil exploded with protests about the country's preparations for the World Cup and the stagnation of services for the nation's citizens. On the way to one of the protests one Tuesday night with my roommate Cláudia, she explained quite succinctly how and why she thought I looked like a Brazilian. As we rode the green line subway toward downtown, she explained: "In Brazil we think that in the US people are either *really white* or *really black*. There your white people are more white than the white people here." She said: "We also think that black people there are really black. Because we are really mixed here." I said that we (African Americans) look just like Brazilians of African descent in the United States. Grasping a concrete example, I told Cláudia that she resembled my mother. My roommate has similar honey brown skin as my mother has and similar features. She asked: "So you look more like your father?" I said yes, I am about the same color as him (a darker brown). As the subway doors opened to our stop, Cláudia turned to me and said, with complete sincerity: "Reighan you are not *pure*. You are not *purely* black." I did not know what else to say other than "I know this" and "it is really a matter of ideology rather than actual phenotype and color."

Cláudia's ideas about race in the United States and Brazil clearly mapped racial features and complexions onto each country's respective populations. It seemed logical to her that if people are not racially mixed in the United States they would be darker or might more closely resemble what we consider a black phenotype. Yet, my brown skin and curly hair troubled this assumption, as does the appearance of many African Americans and many Afro-Brazilians.

Yet, Cláudia was undergoing a process of shifting ideas about race in the United States and in Brazil resulting from her move to the city of São Paulo and her increasing exposure to international influences.

Cláudia was a young professional woman living in the city of São Paulo. She worked in computer systems for a multinational firm in Brazil. As she told me, her father was black and her mother was white, and they migrated south to São Paulo from Salvador about 30 years ago in search of better economic opportunities. Cláudia had honey brown skin, coarse hair (which she chemically straightened), a fleshy nose, and full lips. She referred to her hair, nose, and lips as "traces from her father" (*traços do pai*). She grew up in the rural interior of the state of São Paulo and attended public school her whole life. Her family had enough money to provide the basics: food, clothes, and shelter for her and her five siblings. She was always very studious and managed to pass the vestibular, the college entrance exam, and enter into a public university to study computer systems. She came to the city of São Paulo for the current job that she had. She had little contact with people from the United States, although this was increasing with her economic upward mobility and relationships in her workplace. Many of her colleagues at work had traveled or lived abroad in the United States and in other countries as well. For example, her boss was going on vacation to Tunisia for two weeks. Additionally, everyone in her office followed NBA basketball and she insisted that we watch the NBA finals between Miami and LA. She was rooting for LA and I was for Miami. Coming to the city of São Paulo as well as her interactions with people at her work exposed her to more international influences than she encountered in the rural interior. When I stayed with her, Cláudia was saving money for a month-long trip to the United States in about a year. She was tired of hearing about the United States second hand; she wanted to see it for herself.

Cláudia was a voracious reader, and preferred the genres of fiction and fantasy. She started reading the Portuguese translation of the book *The Girl Who Fell from the Sky* by Heidi Darrow. She told me that the story was about a mixed (*mestiça*) girl named Rachel Morse, the offspring of her white and Danish mother and her African American father. After a family accident in which only Rachel survived she went to live with her African American grandmother. In this new life, she was forced to confront questions of racial identity, which her parents had previously concealed from her and her siblings. Cláudia seemed to choose the book because of conversations we were having about race and definitions of blackness in the United States and Brazil and because the protagonist's identity in the novel generally matched hers, in that they both had white mothers and black fathers. My roommate wondered whether the book was autobiographical at all. A quick Google search took us to the author's page, revealing that Darrow was also the child of a Danish mother and African American father. The featuring of a mixed-race girl and the author's identity as of mixed race troubled Cláudia's ideas about racial purity and racial mixture in the United States. Her previous rendering of the United States as a nation with "really white whites" and "really dark blacks" and of Brazil as the place of mixture was called into question when presented with someone of mixed race in the United States.

Cláudia expressed going through similar issues as the character of Rachel Morse in the book. As someone who attended public schools with large numbers of children of African descent, my roommate told me that the darker children rejected her because she was not dark enough. But the lighter children also made fun of her for being too dark and having coarse hair. This book was the first time she had seen a representation of this dilemma. She said: "Maybe because in Brazil racial mixture is normalized we don't talk about these issues." I just listened as she reflected on her childhood when she had to deal with these issues and hypothesized about why people generally do not talk about this topic in Brazil. She concluded that she "didn't have these issues anymore" and she felt much more secure about herself. The book *The Girl Who Fell from the Sky* mirrored for Cláudia certain experiences that she had had growing up in Brazil. If she had the same experiences in Brazil as were depicted in the novel by an author from the United States, what does that say about racial dynamics and racial identification in these two countries? I am not suggesting that Brazil and the United States are exactly the same when it comes to racial dynamics. But representations of racial dynamics in either country can unsettle ideas about phenotype that are thought to accompany a racially binary United States and a racially mixed Brazil. At the end of my stay in Brazil, Cláudia said: "You know I have really been thinking about a lot of things. You really made me rethink my previous ideas." I do not want to suggest that I was the catalyst of some complete transformation Cláudia experienced. Rather, I accompanied other influences from the United States, such as the book *The Girl Who Fell from the Sky*, which disrupted her previous typology of Brazilian mixture and brownness and US racial purity.

Conclusion

Christina's mother Mariana and her friend, as well as Cláudia, insisted that I looked Brazilian. For them, I fit into the phenotype of brownness that many Brazilians assume to be the national look. Of course, this ideal of brownness obscures the reality that Brazilians fall along the range of colors from black to white. There are instances in the chapter where the presence of African Americans in Brazil was able to disrupt ideas of Brazilian mixture and

US American "pure" racial categories, such as Cláudia reading about racial mixture in the United States. These were moments that exposed Brazilian brownness and US blackness as ideological constructs that have been used to differentiate and categorize multiracial populations in the United States and Brazil—both countries with histories of miscegenation that has produced populations that have every shade of the color continuum. At other times, these ideologies of Brazilian mixture and US purity seemed to remain intact. I learned from my research interlocutors that the nation and its attendant values, beliefs, and narratives can inform the terms of seeing and interpreting others.

The interactions I have described in this chapter took place against the backdrop of the black movement in Brazil. Activists have worked to consolidate the population of African descent under the term black or *Negro*, thus denying the existence of a middle category of mixed race. G. Reginald Daniel describes this project:

> At the turn of the century a small cadre of African Brazilian activists in São Paulo began to articulate a racial ideology that rejected the traditional distinction between blacks and multiracial individuals based on color, instead assuming a racial identity as *negros*. This rearticulation of blackness through a cultural initiative that included mulattoes as well as blacks was similar to the formation of African American identity in the United States and would become a central organizing principle of African Brazilian political discourse through the twentieth century. This radical usage of *negro* sought to imbue the term with not only positive but also political significance. (2006, 78)

Thus the topic of mixture versus purity has great political significance to the black movement in Brazil.

Brazil and the United States have frequently been drawn into relation with each other through the lens of racial ideologies and race relations in each country. Traditionally, Brazilians were thought to exhibit more harmonious race relations due to centuries of miscegenation. In contrast, the United States was cast as the land of racial binaries where those with "one drop of black blood" were defined as black and those without were defined as white. Additionally, the United States was thought to have more hostile race relations and more virulent forms of racism. In this chapter, I have demonstrated how these ideologies convey assumptions about the aesthetic qualities of African-descended people in the United States and Brazil and how these ideologies and assumptions mediate perceptions about who looks and does not look Brazilian. At stake in the maintenance of Brazilian brownness and US blackness is that these ideas can leave intact assumptions about minimal racism in Brazil due to racial mixing. Additionally, assumptions that black phenotypes in Brazil reside in or come from Salvador contain these phenotypes in certain geographic regions, making them exceptional to the norm of Brazilian brownness. Revealing that miscegenation occurred in the United States and that African Americans have white ancestry (among other ethnic ancestry) has the potential to challenge the idea that Brazil is somehow exceptional in its brownness and in its ideal of racial harmony. In other words, if racial mixture and racism can exist in the United States, then miscegenation is not necessarily an antidote to racism and racial inequality. Challenging the assumptions about minimal racism in Brazil is of particular importance as researchers and black movement activists work to expose entrenched forms of racial inequality and demand programs that redress excluded populations (Hanchard 1994; Htun 2004; Martins et al. 2004; Paschel and Sawyer 2008). When people of African descent from different nations or regions encounter one another, they do so across gaps of history, nation, culture, belief, and ideologies. It is important to interrogate the ethnographic contours of such diasporic encounters and the aesthetic beliefs that mediate them in order to reach the moment of realization that despite our geographic distance the challenges of racial inequality remain the same.

Note

1. I use the term African American to refer to people of African descent from the United States and Afro-Brazilian to refer to people of African descent from Brazil. I also use the term black to refer to people of African descent in general. I realize that these terms may obscure as much as they reveal with regard to black peoples and cultures in the Americas.

CHAPTER 8

Changing Notions of Blackness in Field Research in Salvador, Bahia, Brazil

Gladys L. Mitchell-Walthour

In this chapter I examine how my notions of blackness were challenged as a black researcher conducting research in Salvador, Brazil. I have conducted research in Brasília, São Paulo, and Rio de Janeiro but solely focus on Salvador in this chapter. My research spanning a period of 11 years is about the role of racial identification, racism, and group identity on political behavior. In this chapter I focus on the period from 2004 to 2012, during which my research consisted of interviews with politicians about how they address racial issues during campaigns, survey interviews with Afro-Brazilians examining racial identification, racial attitudes, and political behavior, and a survey experiment with Afro-Brazilians studying whether racial and class frames impact support for racial or class policies. I also conducted in-depth interviews with Afro-Brazilians about black identity and affirmative action. In this chapter I do not analyze results from my research. Rather I focus on how I was challenged to analyze blackness more critically because of my lived experience as well as varying notions of blackness I witnessed living in Brazil while I conducted my research. Considering varying notions of blackness within one country is especially important to theories of the African diaspora where oftentimes there is an assumption about common experiences and interpretations of history. The lived experience as a black American researcher in Brazil allowed me to gain a broader understanding of what it means to be an African descendant in another part of the African diaspora. I also began thinking about new ways of analyzing racial politics in the United States when considering black Americans.

What does blackness mean in Brazil and how is it constructed and understood when identifying black foreigners? The country has changed significantly since the first time I traveled there in 2003. It has a more robust economy although currently the economy has dipped. Support for racial policies such as university affirmative action has increased dramatically and today a majority of Brazilians support the policy. While there are physical markers of development such as construction projects and heavier traffic on streets in Salvador, the everyday experiences of race and racialization have not significantly changed for a black researcher. As a researcher I have moved through elite spaces meeting with federal-level politicians in Brasília to low-income black communities in Salvador and São Paulo. In these spaces my identity is negotiated as an American researcher and sometimes denied in favor of identifying me as Brazilian or more recently as Angolan. I discuss the various Brazilian interpretations of my blackness and how I was placed in the Brazilian racial hierarchy and its pigmentocracy when I was mistaken for a Brazilian. I also examine how blackness was used as a marketing tool when I was perceived as an African American tourist by Afro-Brazilian capitalists.

My ideas of Brazilian blackness were challenged when beliefs of a transnational blackness and racial allegiance were manipulated in given contexts and when my skin color was associated with a nation other than my own. These experiences led me to critically question what blackness means and the limits of an analytical category of blackness when examining political opinion in Brazil. My past research examines how politicians use cultural and racial symbols in their campaigns to seek Afro-Brazilian votes and the role of racial identification in supporting black candidates. My current research examines the impact of racism on the political behavior of Afro-Brazilians. This new area of study, which examines racism in political behavior, is largely the result of the interviews I conducted in Salvador, São Paulo, and Rio de Janeiro in 2012 as well as my racialized experiences in Brazil which sometimes mirrored the experiences of middle-class Afro-Brazilians.

Blackness

In Salvador, I identified six notions of blackness which are not mutually exclusive. This is not an exhaustive list but is simply meant to identify some of the types and practices of blackness displayed throughout my time in Brazil. The six types of blackness I identify are a body-centric notion of blackness, a political identity, a locational identity, a commoditized identity, a nonpolitical identity, and an innate identity. Some scholars have made the claim that one commonality connecting African descendants throughout the transatlantic is the history of slavery. Gilroy (1995) claims that African-descended communities throughout the Atlantic have a shared culture. Yet Tanya Golash-Boza notes that a theory of double-consciousness applicable for African descendants through the Atlantic is inappropriate for some African-descended communities. She finds that Afro-Peruvians identify as black but do not acknowledge a shared history of slavery and when they discuss slavery it is in reference to exploitation of labor, which is not limited to blacks. In her research there is no collective memory of slavery restricted to African descendants. The presence of Brazil's history of the enslavement of African descendants is evident in Salvador. In fact, the main tourist site is Pelourinho, which means the whipping post, and the history is ever-present in the city through cultural forms that developed during slavery. Living artifacts of slavery through art forms such as capoeira and representations of candomblé are clearly seen throughout the city. While there is debate whether capoeira was African-born or was created in Brazil (Downey 2005), without a doubt it is packaged and sold as an art form that was practiced by enslaved African descendants and trained and untrained teachers use it as a way to draw tourists to their classes. Blackness in Salvador is clearly seen as cultural but also exists in other forms.

Body-Centric Blackness

One notion of blackness is that it is body-centric (Pinho 2010). Patricia Pinho finds that Ilê Aiyê's notion of blackness is one tied to the body and is innate, that blacks are born with exceptional dancing skills. She also finds that stereotypes about black bodies are accepted by young dancers in the group who often believe they have to make sure they take extra showers and keep themselves clean because blacks have strong body odor. Although this group is often cited for its work of empowerment, lifting the self-esteem of blacks, and embracing blackness and the continent of Africa, many of those involved in the group believe in an essentialized notion of blackness. Similarly, when I conducted research in Brazil, other transmitters of culture also promoted the idea that blackness was linked to the body. There was a belief that blackness "is in the blood." During my first trip to Salvador to study Portuguese, I took private capoeira lessons because I was too embarrassed to join a group class. My capoeira teacher was encouraging and insisted that as a black person capoeira was "in my blood." I was not convinced as it is a very difficult sport and when I was forced to enter the Roda on a Friday night after only a few classes, I was convinced that capoeira was "not in my blood." In the Roda I had no idea what to do and when I finished someone asked me: "Well why didn't you do this move or that move?" I wondered whether my capoeira master (*mestre*) genuinely believed that I could become a good capoeirista because of an innate athleticism or if he was using blackness as a way to sell his teaching services.

Blackness as a Political Identity

Blackness is also a political identity, which I found evidence of when analyzing my survey interview responses. For example, I found that Afro-Brazilians who identify as black rather than nonblack, such as *moreno, pardo,* or *moreno claro,* are more likely to vote for black political candidates. Blackness as a political identity was evident when I saw young hip-hop artists performing songs that embraced blackness. During my fieldwork in 2005, I attended a hip-hop show where one group's refrain was *Nós Somos Pretos*! We are Black! This greatly impressed me as I saw young people embracing blackness, which was very different from France Twine's (1998) work on a rural community outside of Rio de Janeiro and Michael Hanchard's (1994) work in Rio de Janeiro and São Paulo, where Afro-Brazilians did not embrace blackness and were not willing to become involved in black activism because of the belief in racial democracy. In 2008 and 2012, when I interviewed Afro-Brazilians, including activists and nonactivists in Salvador about why they identified as black, some interviewees claimed it was because they were proud of their ancestry or history.

Yet the majority identified as black simply because of physical traits and noted that blackness was not something they could escape. This is in sharp contrast to the supposed ambiguity of race in Brazil. Derek Pardue (2004) also finds that hip-hop artists in São Paulo address racial discrimination in some of their songs. Similarly these artists in Salvador addressed the issue in their performance. Ilê Aiyê was another group that politicized blackness as a means of empowerment and pride. Although, as mentioned earlier, some notions of blackness are body-centric, the group often discusses the beauty of blacks in their songs. I attended some of their performances and noticed this common theme as well as the prominent role dark-skinned dancers played in their group.

Locational Blackness

Another idea of blackness is that it is located or situated in certain spaces. When living in the neighborhood of Rio Vermelho, a close friend of mine who was light-skinned and identified as *parda* would often say I looked Brazilian. One day she remarked: "You look like you would live in Liberdade!" Liberdade is a low-income neighborhood where Ilê Aiyê's school is located. My friend happily made the statement and did not say it out of malice. There were clearly Afro-Brazilians living in the middle-class neighborhood of Rio Vermelho, yet because of my dark skin I did not belong in this neighborhood but in another where dark skin and low-income status go hand in hand. Similarly, in Cecilia McCallum's work in Salvador, she notes that public spaces are marked as black such as the city bus at certain times when it is primarily used by domestic workers.

Nonpolitical Blackness

It is important to note that blackness can also be nonpolitical. In my in-depth analysis, some people identify as *Negro*, the racial category for black, but may not have progressive stances on racial policies such as affirmative action. In general, in my interviews respondents with more conservative views on racial policies were those with less education. In daily life, nonpolitical blackness is manifested as a superficial category that may be in fashion so people may embrace an aesthetic blackness without actually holding progressive political opinions or without a sense of group attachment. The notion of a nonpolitical blackness was made clear to me when I attended the play *Cabaré da Raça*! At different moments during the play one of the actresses would shout *Negro é lindo* (Black is Beautiful)! At other times, they would discuss how black was in style or in fashion (*o negro está na moda*). The play articulated the way I felt in a city where blackness was used as a fashion statement in terms of women wearing thick braids or afros and young men wearing cornrowed hair or natural and faux locs, yet this superficial pride has not translated into political and economic power for blacks. Some of my frustration grew with the fact that, in terms of politics, Salvador has never elected a black mayor. Olívia Santana is now the Vice-Mayor and all the candidates for Vice-Mayor were Afro-Brazilian in 2012. However, this only reaffirms the racial patriarchal status of blacks in Salvador. They can be placed in a second position of power but not in the first as leaders of the city.

The Commodification of Blackness

Anyone who spends significant time in Salvador will see the commodification of blackness. It is packaged and commodified for tourists. This soon became apparent as I lived in Barra in 2004 and briefly again in 2005 before moving to Rio Vermelho. I noticed out-of-shape men doing a couple of flips in an attempt to pass themselves off as capoeiristas and asking tourists to pay them money. Two particular stories come to mind. Some of these men claiming to be capoeiristas hung out near Porto da Barra and Pelourinho. One night I hung out with some Americans, both black and white. As we were walking through the main square some "capoeiristas" performed and then asked for money from the whites. They did not ask any of the African Americans in the group for money, presumably because they assumed we were Afro-Brazilians. The fact that we were not asked for money but the whites were frustrated some of them. However, in this case blackness was associated with "blackness" and perhaps an inability to donate money, while whiteness was associated with foreignness, tourism, and as a potential client for the "sale" of capoeira. Watching both enchanted tourists and guys spinning and doing jumps disturbed me because whites were the onlookers and blacks were there to sell their athleticism for entertainment.

The second story I recall demonstrated that everyday citizens can take advantage of such a rich art form to make a few dollars. I met an African American woman from New York who had moved to Atlanta. She was in her early twenties and was a free spirit. She had traveled alone to Salvador to experience the culture and figured that with her positive energy she would be able to navigate the city safely. She lived in a hostel in Barra and one day while walking along the beach watched some "capoeira" players. She told me one of the capoeiristas offered to teach her so she took him up on his offer as she was interested in learning the art form. She began lessons and noticed he was really flirtatious and he asked her out on a date. She went to the movies with him but was highly upset when he took her to see pornography. While she was interested in meeting Afro-Brazilians as fellow comrades she was disappointed that he viewed her as a "loose American girl" even though she was African American. I had warned her about people posing as capoeiristas who take advantage of foreigners when she told me she was taking lessons. I advised her to be careful as Barra could be a dangerous neighborhood because people are often looking for tourists. Erica Lorraine Williams (2013) finds that not all Brazilians looking to interact with foreigners have bad intentions and in fact foreigners may also have questionable intentions which lead to ambiguous encounters for both parties. The young lady later stayed in my apartment a couple of nights as someone crawled through the window into her hostel and robbed her. The thief was never caught.

Going to Pelourinho with my American friends became dreadful because it exemplified the commodification of blackness. While there were African descendants there as tourists, I began to simply see whites as buyers and blacks as those engaged in the commodification of and selling of blackness. Even more disturbing and heart-wrenching were the black children and adults who begged. I began to see pure misery walking around Pelourinho. Poor blacks lived in misery while happy white Europeans and Americans walked around clueless of the history of the place or how they were reinscribing power dynamics from centuries ago. It was sickening to me. It was difficult as a black researcher to see how blacks had to play the game of dressing up as "traditional Bahianas," while tourists took pictures with them or older black women who often looked tired but tried to maintain an energetic attitude selling *acarajé* at night to white tourists as a means of survival.

Inescapable Blackness

In Brazil blackness is tied to physical traits and ancestry. Robin Sheriff (2001) finds that in a slum in Rio de Janeiro, most Afro-Brazilians have a dichotomous way of thinking about race: one is either black or white. This is similar in Salvador where racially most people view blackness as an innate identity that one cannot escape. In the in-depth interviews I conducted in Salvador the common response to my question "Why do you identify as black?" was that they were born that way and people referred to physical attributes such as a "flat" nose or big lips, or mentioned curly (*crespo*) hair. Similar to the Afro-Brazilian respondents, people in Salvador identified me according to physical attributes. On my last research trip in 2012 my nationality was most often misidentified. However, it is not an uncommon experience for middle-class Afro-Brazilians to be misidentified as African American in certain settings.

During my 2012 trip, on a number of occasions in Salvador and São Paulo people assumed I was Angolan. In Salvador, the white bank agents insisted I was Angolan even after I showed my US passport. I went to the bank Itau to pick up money from my colleague to pay our research assistant. The bank agent would not let me receive the money unless I showed some type of proof of where I was living in Salvador. I told her I was staying with a friend so I did not have any documentation. I showed my US passport and driver's license but she would not accept these forms of identification. I asked to speak with the manager (*gerente*) and she said the same thing. While explaining the situation to me, she clarified the money arrangement saying "your friend from Angola is sending the money." I laughed and asked, Angola? No, he's sending it from the United States. My frustration was due to a misidentification of my nationality and the fact that blackness is not simply located in two places—the African continent and Brazil. Although I know that Brazil has the largest number of people of African descent outside of Nigeria, such dichotomous thinking about the location of blacks erases African descendants in other parts of the diaspora. As an African American I identify as black or African American. I acknowledge I am of African descent and am aware of the history of slavery and resistance to the institution and present-day manifestations of white supremacy in this country. In a sense, I felt that my history was erased as my nationality was erased. Despite the president of the United States, Barack Obama, self-identifying as an African American and making blackness visible on an international stage, and despite the fact that I had a US passport, the people working in this bank felt an obligation to not acknowledge me as a black American but as an Angolan. In this way for these white *soteropolitanos* their investment in their conception of blackness hindered them from equating foreignness and blackness with an economically developed country. Black foreignness was linked to Angola, a developing country.

Tanya Golash-Boza (2012) finds that Veracruzanos of African descent in Mexico do not acknowledge their African ancestry and attribute their dark skin to the sun. They often view blackness as something foreign. In this case, blacks are identified as Cuban or American but not as Mexican. For much of Mexico's history they upheld the national ideology that their racial roots were made up of indigenous people and Spaniards. Unlike Mexico, Brazil has always acknowledged its racial history as coming from Africans, Portuguese, and indigenous people. Yet in Brazil, often when African culture or African nations are discussed, it is in a folkloric sense. Africans are viewed as people who contributed to culture through dance, music, and food. Thus they are relegated to contributing to culture historically and often frozen in time and viewed as primitive. Such an understanding of blackness and Africanness as backward and underdeveloped is a common belief for many Brazilians. It is in this context that I was viewed as an Angolan.

Diaspora and Blackness

In his early work, Abdias do Nascimento, a leading black activist, advocated a notion that would connect blacks throughout the diaspora. In his book *Brazil: Mixture or Massacre* (1989), he was concerned with the marginalization of blacks and how blacks could unite and resist oppression. Although this writing reflects the concern of the time with efforts of decolonization in African countries, his notion of pan-Africanism is essential in thinking about the African diaspora in the contemporary context. Nascimento argued that culture is not static but rather flexible. He believed that African culture calls for unity and self-reliance. It is with this concept of self-reliance in a collective sense that Nascimento believes blacks throughout the diaspora can write about their particular history as black people and can rely on "African economic forms" to support their communities (28). What I find most important in terms of diaspora is Nascimento's focus on resistance to oppression and marginalization of blacks in Africa and the diaspora. For Nascimento, this was a key aspect of defining the African diaspora.

As a black American researcher in Brazil I witnessed and experienced racism but also saw efforts to resist marginalization. Blackness is not just a category denoting marginalization. It also entails resistance to this marginalization. Similarly, in Patricia Hill Collins's notion of the matrix of domination that black women's multiple identities lead to a particular type of oppression, one must understand her theory of empowerment and resistance to the assigned roles society gives African-descended women. While I experienced and witnessed various forms of blackness in Salvador I also witnessed resistance and in my own way resisted by continually frequenting spaces such as restaurants or even clothing stores that were unofficially reserved for whites.

Viewing Blackness in the United States through a Soteropolitano Lens

I critically reinterpreted blackness while conducting research in Brazil because of the various ways in which my own blackness was interpreted, as well as the various manifestations of blackness I confronted in Salvador. My notion of blackness as a political identity was challenged, yet it aided me in developing a more nuanced account of blackness in Brazil and the United States. In the United States, many people refer to the "one drop rule" as a way of defining blackness despite ongoing research to the contrary (Rockquemore and Arend 2002). Biracial Americans and Latinos of African descent may self-identify in a number of ways depending on the context. Thus their identity is not static and is ambiguous. In the United States, blackness is also sometimes simply viewed as biological and often in research on black politics while some scholars offer a rich account of the historical legacy of racism against blacks they do not actually interrogate the meaning of blackness in statistical analysis where race is simply a dummy variable. Rather than assume *all* people of African descent interpret their experiences in the same way, scholars of black politics must explore the various notions of blackness in the American context.

Conclusion

As a black researcher I have experienced Brazil in a particular way because of my skin color and physical features. All scholars doing work in Brazil, regardless of skin color, will have experiences based on their various intersectional identities. Some may be treated in a very privileged manner while others may not. As I think about the way my research has developed, it is significantly shaped by how I relate to the stories and experiences my respondents have recounted. When middle-class Afro-Brazilians recount experiences of racism it is very difficult for me to make the traditional claim that scholars have made that one's racial or color identification depends on one's status. This is simply not the case. Such claims should be understood in terms of what Sheriff (2001) witnessed in her fieldwork where people's colors were whitened among friends and acquaintances in an effort to be polite. However, when one is in the social world where no one knows you, you are judged and treated according to physical attributes. This was certainly my experience in the country.

Having an embodied experience of blackness does not limit the way in which one does research. In fact, it allows for a more nuanced approach to one's work. Keisha-Khan Perry (2013) experienced many of the same racist encounters Afro-Brazilian women faced. This was confirmed by black women organizers. Rather than a distant and unattached approach to research, Perry was embedded in the community work of her research respondents. This allowed for a deeper analytical approach to research on black women organizers. It is the responsibility of African descendants throughout the diaspora to consider how they are interpreted in the countries in which they conduct research and to critically evaluate and reevaluate their research according to these contexts.

CHAPTER 9

An African/Nigerian-American Studying Black–White Couples in Los Angeles and Rio de Janeiro

Chinyere Osuji

Why Interracial Couples?

I moved to Los Angeles from Chicago in order to work with Edward Telles in the Sociology department at the University of California, Los Angeles (UCLA). I knew that I wanted to study race relations in Brazil and would probably do research on affirmative action policies that had recently been implemented there targeting the poor as well as people of African descent. I had never been to Brazil, but was interested in the African diaspora in Latin America for a long time and had studied Portuguese for a semester at Harvard. I spent my first year at UCLA fulfilling Sociology department requirements as well as using my fluency in Spanish to learn Portuguese.

At the end of my first year, I went to Brazil for the first time, and spent my time working at Geledés Institute for Black Women (Instituto da Mulher Negra), an NGO specializing in black women's issues in São Paulo. The few interracial[1] couples I saw during this trip struck me. I was expecting couples with a random color assortment, but instead I saw that couples largely shared the same skin tone, regardless of what that color was. I saw far more cross-color or interracial couples in Los Angeles than I had in São Paulo! In Los Angeles, my black girlfriends and I had commented on how we often saw black men with white, Latina, or Asian women, while it was rare to see black–black couples or black women paired off with *anyone* in Los Angeles. Several of us had the sense that black women were left out of the dating market. Judging from the empirical literature on black women's limited dating and marriage in the United States, we were right on the money.[2]

In São Paulo, I lived in the ritzy Jardins neighborhood where the only people who did not look like they were solely descendants of Italian and Portuguese immigrants were domestics, chauffeurs, and security guards. I felt that I stood out far less in Westwood, West Los Angeles, and Santa Monica than in Jardins. In addition, notwithstanding the large proportion of Asians in São Paulo, I rarely saw them interracially partnered off, if I saw them at all in Jardins. Despite moving to a country where whites only made up half of the population, and in a city where Afro-Brazilians were approximately 35 percent of the population, I lived in a predominantly white world. With most Afro-Brazilians relegated to the lowest rungs of the socioeconomic ladder, class segregation aided the racial segregation I experienced firsthand.

This was not supposed to happen. Brazil is a country that has celebrated its history of race mixture and characterized itself as a racial democracy. In this ideology, popularized by Gilberto Freyre, color is not supposed to be an impediment to interpersonal relations.[3] However, preventing relations between black men and white women was part of the justification for *de jure* racial segregation in the United States. Yet, I had seen far more interracial couples in Los Angeles than I had in São Paulo. Something was amiss and I was determined to find out what it was.

After my return to Los Angeles, I talked to Telles about these issues and decided to do a comparative study of black–white couples in Brazil and the United States, with Rio de Janeiro and Los Angeles as my research sites. Before defending my dissertation proposal, I went to Brazil once more, to conduct preliminary interviews with black–white couples. This time, I went to Rio de Janeiro, where it would presumably be easier to find couples to interview. Half of the city identifies itself as black or brown, both often collapsed into one large Afro-Brazilian population, and the other half as white. Between the numbers that should have worked in my favor, my contacts in Rio de Janeiro and those of Eddie, I was bound to find a subset of the couples that Brazilian demographers would identify as interracial couples.

Black Men, Interracial Marriage, and the Black Movement

With the transition to democracy in the 1980s, the black movement in Brazil increased its presence. The movement's exertion of political pressure has coexisted with increasing rates of interracial marriage in Brazil. On a number of occasions, individuals in Rio de Janeiro told me that interracial couples involving black men with white women were very common in the Brazilian black movement. In fact, some Brazilian black female activists have discussed its prevalence.[4] Although interracial relationships were something that people affiliated with the black movement discussed among themselves, it was often taboo to discuss it out in the open. Ironically, when I spoke to black–white couples in Rio de Janeiro, many of them had the impression that the black movement was against interracial marriages and advocated the notion that blacks should only be married to other blacks. The few black men and women whom I spoke with who were affiliated with the black movement discussed the stigma and constraint they felt due to their relationships with white spouses.

I had a few encounters in which I discussed my research with Afro-Brazilians affiliated with the movement. On one occasion, I met up with one of my *negro* informants, Roberto,[5] to attend an event at the State University of Rio de Janeiro (UERJ). We were going to hear Abdias do Nascimento speak, the now-deceased black movement activist and scholar. As we waited for the event to commence in front of the auditorium entrance, I saw several people with various shades of brown skin. Initially, none of them looked like they might have been the descendants of Portuguese or Italian immigrants with their lighter skin and straighter hair. However, I then saw a man with caramel-colored skin and dark-colored pants hugging a slender, blonde woman with very light skin. I wanted to ask Roberto about them, but we ran into one of his friends, Marcos, a sandy-colored man wearing a dashiki. Roberto told him that I was in Brazil studying interracial couples.

Marcos asked me: "What do you think of interracial marriages?"

"What do *you* think?" I responded. "I'm the American here trying to study these relationships."

"It is very touchy," he replied, using the Portuguese term *delicado*.

Just then, a chocolate-colored man with long dreadlocks walked over to the three of us. He introduced himself as Pedro. I introduced myself and Roberto explained my research in Brazil. Pedro commented: "In the black movement, the early leaders almost always dated white women." He pointed in the direction of the auditorium and said even Abdias do Nascimento is married to a white woman. "However," he continued, "this is not the case anymore." Marcos quickly chimed in: "You still see it though." He then pointed to my left. We all turned to look in the direction he was pointing at. There was the caramel-colored man that I had seen earlier, walking away from us toward a large bulletin board alone. Apparently, I was not the only one who had noticed him with the white woman.

Pedro suggested that we all sit down at a table nearby. After pulling chairs together to accommodate all of us, Marcos suggested that I sit opposite them. Though I felt slightly uncomfortable with the interrogation-style arrangement, I agreed. Pedro proceeded to tell us about his father's perspective on relationships with white women. "My father told me that I would never have a serious relationship with a white woman. To hook up (*ficar*) is one thing, but having a serious relationship is another." Pedro's father had told him that only the "fat, ugly ones that white men did not want" were the white women available to black men. "I once dated a white woman," Pedro continued, "but my wife is a black woman (*negra*), just a little lighter than me. She had been married to a white man (*branco*) but they broke up and didn't have any kids together." The tone of his voice suggested that Pedro was grateful for the lack of children from this previous relationship.

Marcos then shared how he was married to a black woman and that people often see them as a model couple, since they both have a strong black consciousness. However, this makes him uncomfortable, since, if they were to break up, he worried that outsiders would pass judgment on them. Then Marcos proceeded to tell the story of a white woman that he had a crush on when he was in school in his early twenties. "I liked her, but for reasons of self-esteem, I didn't think that she would like me." Despite his insecurity, he started to "chat her up" (*bater um papo*) and flirt with her. Apparently, his efforts worked. A friend of hers served as a go-between and told Marcos that her friend was interested in him. Marcos smiled as he looked down: "But I didn't think it was possible. I thought she was joking—like, 'What do you mean she likes me? Is this some sort of a joke?' However, it was not a joke, so they started dating. As a consequence, Marcos said that people in school would pat him on the back and say: "Awesome!," "Good job!," and "Congratulations!"

Just then, Pedro cut off Marcos: "I used to go out with a white woman and you know what they said to me?" I shook my head, no. "Congratulations!" he said, tapping my right arm.

Marcos's and Pedro's experiences reveal a variety of meanings surrounding interracial romantic relationships that sometimes were in competition with one another. Black (male) leaders in the Brazilian black movement have been able to have and maintain relationships with white women. In addition, both of these black men show that they, as well as other black people they know, have had relationships with whites in the past. This is different from US civil rights and black power movements in which many blacks openly stigmatized black men for "talking black and sleeping white."[6] Nevertheless (or perhaps as a consequence), Marcos's marriage to his black wife was a positive symbol for other blacks affiliated with the black movement. Outside of the black movement, both Pedro and Marcos revealed that they understood their past relationships with white women being read as great *personal* achievements. These men understood outsiders valorizing them and their romantic relationships according to the color of their female partners.

Previous scholars have shown how whites stigmatize interracial intimacy across the Americas,[7] while others demonstrate how race mixture is revered, including by people of African descent.[8] Black men who date and marry white women supposedly suffer from low self-esteem or have desires to whiten themselves.[9] In the United States, they are also supposed to endure an elevation in gender status through relationships with white women that involve more traditional gender roles than they think they would experience with black women.[10] However, the ways that black men understand and negotiate these meanings in their own romantic relationships has not been adequately explored. My research in Rio de Janeiro revealed an ambivalence of both stigma and esteem for blacks who engage in romantic relationships with whites in contemporary Brazil.

Racial Preferences in Rio de Janeiro

While conducting preliminary interviews, my goal was to understand how Brazilians conceptualize race mixture. I often heard the phrases "blondes (read as white women writ large) love big black men"[11] and "foreigners (often read as white foreign men) love mulatto women."[12] These feelings were often understood to be mutual, with blacks also preferring white partners. Consequently, in both sites, I was supposed to have a racial preference for dating white men. Combined with the hypersexualization of black women in both societies,[13] this led to some uncomfortable encounters with white men.

I remember sitting on the beach with my tanned, brunette, Italian friend, Teresa. Teresa often self-professed her love for big black men or *negão* and described the stigma she experienced when she lived in California because of her racial preference. She was dating Zezinho, an Afro-Brazilian man with medium-brown skin. One day, we were sitting on the beach with three of her girlfriends, all of them with varying tones of brown who self-identified as black women (*negras*). Upon hearing what I was doing in Brazil, her friend Ursula complained about how black men in Brazil do not like to date black women. The two other black women looked at the sand silently as she spoke. Then, one of them, Natalia, a woman with light brown skin and her hair wrapped in a scarf, revealed that she doesn't like black men and only dates white men, preferably foreigners (*gringos*[14]). The last woman, Ana, revealed that she was in a relationship with a white man.

I quickly found out that although black women were having similar conversations to the ones I had in Los Angeles, there was an interesting twist of overt different-race preferences that I had rarely encountered before. As people would say back in the United States, all but one of these women were "down with the swirl," preferring interracial relationships. Unlike the small segment of Afro-Brazilians associated with the black movement, interracial preferences seemed par for the course among another segment of people in Brazil.

Experiences in Los Angeles and Rio de Janeiro demonstrated that something different was happening in the dating markets. Namely, racial preferences in dating were very explicit in Rio de Janeiro, whereas they were more stigmatized in Los Angeles. During my stay in Los Angeles, I had not met a single black woman who had expressed a preference for white men to me, including white foreigners.[15] While I had met black women who had dated white men, including myself, none of us expressed a preference for them. If anything, we had same-race preferences that seemed challenged by the context of out-partnering of black men.

In addition, while white women openly liked black men (*gostar de negão*) in Rio de Janeiro, similar to Teresa, I had never heard white women openly talk about themselves in that way in the United States, even the few I knew who dated black men exclusively. I found this pattern repeatedly in the interviews that I conducted with Brazilian and US white women married to black men. I had never heard about a similar *gostar de negra* among white men whom I met or interviewed in Brazil. Either it did not exist as a common catchphrase or it was taboo to discuss this with a black female social scientist from the United States, like myself. While I did hear men of all colors describe their preferences for *morenas*, the term *morena* is rich with racial ambiguity, so that it can mean anything from a white brunette, a woman of mixed-race ancestry, or a black woman in general.[16]

The sole occasion that I heard a white man talk about himself "liking black women" was in Los Angeles, at a multiracial place of worship. I was introduced to two black women, both East African immigrants, as well as their fellow white male congregant. I explained my hunt for black–white couples to interview in Los Angeles and uncomfortably watched the white man say "I love black women!" as he hugged one of the women, who smiled awkwardly. This sentiment was not something I had ever heard a white man express prior to or since that one occasion.

Finding Couples in Rio de Janeiro

Despite the ideology of racial democracy, in which Brazilians interact freely among one another, empirically speaking, marriages across colors are still in the minority for Brazilians.[17] They comprise approximately a third of all marriages in Brazil. When Brazilians do marry a person of another color, they often marry someone in proximate color categories, with whites more likely to marry people who are brown (*pardo*) rather than those who are black (*preto*[18]). In addition, demographic studies also suggest that many brown individuals are married to blacks. However, the ambiguity of the brown category means that some brown–black marriages include those among Afro-Brazilians, who probably do not see themselves as crossing racial or color lines in marriage.

The United States and Brazil have different understandings of whiteness and blackness, making it often difficult to determine what a black–white couple in Brazil is. In the United States, whiteness is understood as exclusive and is based on perceived (though not actual) ancestral racial purity.[19] In Brazil, black ancestry does not prohibit whiteness, because blackness is determined by phenotype rather than solely ancestry.[20] Many Brazilians fall in a continuum of race, complicating even further what is meant as "black" or "white" in Rio de Janeiro. Nevertheless, as I have argued elsewhere,[21] there is still considerable overlap in how racialization in both societies operates, including racial hierarchies privileging whiteness over blackness. Discussions of racial ambiguity in Brazil tend to overlook that fact.

To find respondents, I followed the example of Brazilian scholars who had done qualitative studies of black–white couples[22] and recruited couples involving a black person (*negro*) married to a white person (*branco*). I asked only native Brazilians for referrals of black–white couples and my requisite was that partners be identified by other Brazilians as *negro* or *branco*. I often told Brazilians that I was looking for couples involving a *branco com negro*, or a white person with a black person, across gender–race combinations.

Despite knowing the demographic literature on interracial marriages in Brazil, I was still surprised at how few Brazilians actually knew couples who fit this criterion. The black men whom I had met at the Abdias do Nascimento talk and even the women with explicit racial preferences in dating had difficulty introducing me to married black–white couples. Yet, I managed to rely on the eyes of other native Brazilians to find 25 couples to interview in Rio de Janeiro, involving both black men with white women as well as black women with white men. The overwhelming majority of respondents (46 of 49[23]) identified themselves in the same ways that their referrals did, similar to nationally representative studies in Brazil.[24] This recruitment process enabled local understandings of race and contemporary race mixture to prevail while providing homogeneity in how outsiders identified and treated black–white couples in both Rio de Janeiro and Los Angeles, while allowing me to examine their experiences from their own perspective.

From "Angry Black Woman" to "Available Black Woman"

In Cooley's notion of the "looking-glass self," through social interaction, individuals imagine their appearance in the eyes of the people they interact with.[25] Subsequently, they incorporate the judgment that outsiders have of them into their future actions with those people, in part to aid communication. Similarly, Du Bois describes how African Americans use this "double-consciousness" to negotiate their interactions with whites in the United States.[26]

As a US scholar, my ability to discern how others saw me in the two settings was fraught with complications. In the United States, the "angry black woman" is supposed to be hostile to black men who are dating or marrying white women.[27] In Los Angeles, I was supposed to be the "angry black woman" trying to understand why white women "are trying to take all the good black men"[28]—in other words, a more livid version of Ursula. For example, my first US interviewee, Allison, was a white woman who revealed that she was "scared" that I was going to interview her. Although I tried to probe why she felt this way, she could not identify the reason for those sentiments. In my many years of conducting qualitative interviews for different projects, that was the first time I had ever had such a response. In addition, since most studies of interracial couples involve white women, prior scholarship on this topic had not adequately prepared me for this image being reflected back onto me.

To overcompensate for this stereotype when interviewing couples in Los Angeles, building rapport with couples, especially white women in the relationship, was of tantamount importance. This was especially important since I would be interviewing husbands and wives separately and did not want to be perceived as jealous of their relationship as an "angry black woman" or as a potential adulterer. I usually approached wives first as a way of letting them know that their permission was necessary and most important for their husbands' participation. In addition, I typically conducted interviews in their homes so that they would be comfortable with the subject matter. Due to my own double-consciousness, I purposefully smiled even more than I normally do to cultivate a nonthreatening, cheery persona that would make couples comfortable and aid communication. In addition, I started interviews by asking respondents about their childhood, their experiences in school, and their parents' occupations while their children were growing up. This enabled me to build a rapport with the respondents before asking them difficult questions about their dating histories or their racial preferences in dating and marriage. I was very conscious of smiling and nodding during these moments to make respondents feel comfortable revealing potentially taboo sentiments.

However, in Brazil, I was constantly wiping away the fog from my "looking-glass self"; I could not always perceive or understand how other people saw me as a black woman studying black–white couples. Like other scholars before me, I saw myself as an inquisitive social scientist of color concerned with the condition of racialized minorities.[29] As a result of this prior scholarship, I was mostly prepared to negotiate my identification as an African American, a US citizen, and a woman. However, I was not prepared to be seen as a black woman who was sexually available, particularly to white men. Since I was not Brazilian, I had thought that I would be immune to this notion.

For example, I had an uncomfortable encounter with a taxi driver in Salvador, Bahia, who was taking me to a hotel. With his trunk completely full, my entire luggage found space in the back seat. I sat up front, next to the driver, as I had done many times when I lived in Spain. The driver was an older white man with white hair, peach-colored skin tanned from the sun, black eyebrows, and a black mustache. I told him that I would soon be in Rio studying interracial couples (*casais interraciais*). He informed me: "I have had many girlfriends who were *negras*." I saw the gold ring on his left hand and was excited that I had possibly found a couple with whom to practice interviewing. I asked him if he was married. "No, no," he replied. My curiosity turned to confusion as I realized there were two possible scenarios: (1) he might be pretending he was not married to hit on me *or* (2) perhaps Brazil was like many places in Europe where they place their rings on the right hand, not the left.

> It was a very awkward moment. I asked him a second time: "You said that you've had many girlfriends who were black (*negras*), right?"
> "Yes."
> "Are you married to a black woman?" I inquired.
> "No, she's not black."
> "So, what is her color?"
> "A little darker than I am."
> I thought about how color is relative in Brazil, with a person sometimes describing their color in comparison to those around them, such as proximate friends and relatives.[30] For this reason, I asked: "So, it's because she is darker than you?"
> "She is *morena*."
> Fully aware of the ambiguity of the term *morena*, I played the dumb foreigner or *gringa*. "Oh, she is *morena*. What does *morena* mean?"
> "She is mulatto (*mulata*)," he replied.
> "Oh, she is *mulata*," I repeated.
> "Yes, because in Brazil, white and black make *mulato*." He then proceeded to explain how Brazil has indigenous people and when you mix them with white, you get a mixed person (*mestiço*). He stressed that Brazil is a multiracial society with people from a variety of different races.[31]

At this point, the mosquito bites I had gotten the night before started itching. I looked down at my thigh to inspect the bites and the taxi driver looked down to see what I was examining. I felt strange, because I did not want him looking at my thigh. He proceeded to ask me: "Why don't you stay in Brazil?" and "a *morenona*[32] as pretty as you would have a lot of success here in Brazil."

I told him that I wanted to be a university professor and he tried to convince me that it is also possible to do that in Brazil. I decided that it was time to change the subject.

I asked him if he knew any jokes about blacks and he replied: "Yeah. Many people think that black men (*negros*) are like this." He took both of his hands off of the steering wheel, and placed his palms parallel to one another about a foot apart. It was a reference to black men having large penises. I did not appreciate the humor, but I hid my surprise and nonchalantly asked if he knew any jokes about black women. He said that many people think that they are "hot in bed." We soon arrived at the destination, abruptly ending our conversation about stereotypes of blacks, specifically black sexuality.

In another incident, I was hanging out with Teresa, Ursula, Natalia, and Ana, during a party for "couch surfers" that Teresa had been invited to. Couch surfing is a phenomenon in which adventurous people, usually from Western countries, travel the world through a network of people who allow them to sleep on their couches, at little to no cost. A network of couch surfers had decided to have a beach party, complete with fire lit torches and a guitar-playing, white, 20-something man. The couch surfers were drinking cans of beer 15 feet from the water. Teresa, taking in the predominantly white scene, expressed disappointment at the lack of Afro-Brazilians at the party and the "lame" guitar strumming that reminded her of beach parties back home in Italy.

Nevertheless, together, we approached the group of "couch surfers" on the beach. We met three men who were part of a group. Eduardo was tall with milky skin and very short, dark brown hair. He had dark brown eyes and wore glasses. His friend, Gustavo, was a slight man with medium-brown skin (perhaps a *moreno claro*) with hair that was cut close to his head, almost shaved. Their other friend, Alexandre, was a chubby medium-brown black man with brown skin and shiny black eyes. They were all much taller than my 5'3, so I felt even more out of place.

Eduardo asked me: "Where are you from?"
"I can't tell you," I responded. "If I told you, I would have to kill you." Everyone laughed at my oft-repeated joke.
Gustavo said: "You must work for the CIA or something."
"Yes, it's true. So watch out!" I replied.
After taking guesses at one another's ages, Eduardo asked me: "What are you doing in Brazil?"
"I am studying interracial marriage in Brazil," I replied.
"Blacks and whites (*brancos e negros*)?" he asked.[33]
Once more, I said: "I don't know. You tell me—I don't know. I'm the American here trying to understand what an interracial marriage is here in Brazil."

"It's very complicated," Eduardo continued.

"Why is it complicated?"

"Because everyone is so mixed."

"So, what does that mean?" I asked.

"For example," he explained, "the mother of my mother was indigenous and the father of my mother was Spanish. The father of my father was Italian and the mother of my father was Spanish."

"So the mother of your mother and the father of your mother—were they an interracial marriage?" I asked.

"Yes."

"Those are the types of marriages that I am trying to study, but not in the past, today."

"But that's difficult because," he said as he brought Eleanor to stand next to him, "for example, she's here and she has white ancestry. Many Brazilians, no matter what their color, have black ancestry."

"So, if you two were to get together, would you be an interracial couple?" I asked. Eduardo looked uncomfortable and said nothing. There was an awkward pause in the conversation, so in my attempt at being a good ethnographer, I waited to see what would happen.

Eleanor triumphantly said: "I'm a black woman (*negra*)." This was followed by another awkward pause.

Looking at both of them, I decided to ask the question again: "So if you two were to get together, would you be an interracial couple?"

Another uncomfortable silence followed. He said nothing, she said nothing, and I said nothing.

I sensed that he did not want to offend her by saying that she was black, despite the fact that she self-identified that way. Alternatively, perhaps, he didn't want to place a racial label on her because, oftentimes in Brazil, to identify a person's race is being racist. Another possibility is that by saying that they are an interracial couple, Eduardo would be denying the race mixing that has already taken place in their ancestries. Also, perhaps he was uncomfortable with identifying himself as white, especially to a North American who lived in the country of the Anglo-Saxon "true whites."[34]

I could no longer stand the awkwardness. "Well, that's what I'm studying," I said. "Do you know any interracial couples?"

He thought and he thought and he thought some more. Finally, after a long pause, he said: "Yes I know one." I was surprised at this response, given narratives I had heard of race mixing being common in Brazil. I assumed everyone, including him, would know several couples and that race mixing must still be happening to a great extent in Rio de Janeiro. We quickly changed the topic.

As we conversed, Eduardo laughed at something I said and suddenly grabbed me close to him, pressing my lips against his neck. I was flabbergasted and disgusted at the same time. I maintained my distance from him for the rest of the night. Somehow, he had misconstrued my comments as flirtation or an open invitation to affection. More likely, being a black woman meant that I was available, whether romantically or sexually, and I was not taken just for the curious, quirky social scientist that I thought I was. I decided not to pursue the couple he had recommended in order to avoid further contact with him.

I suspect that studying interracial couples in Brazil made me seem like a black woman in search of her white boyfriend/husband/hookup. However, I find it striking that I did not have awkward experiences with Afro-Brazilian men. It is possible that they felt the same way; I just never had interactions that suggested this was the case. This may have been due to my spending free time with Afro-Brazilians who had a strong *negro* identity and saw my research on race as a more serious endeavor.

Many of my social identities and statuses are not available to many women who look like me in Brazil, yet some social categories, such as race and gender, are stickier than others. I became particularly aware of this while waiting for a bus one night in the middle of Rio de Janeiro. A white man in a black suit, black hat, and curly sideburns approached me in what I took for an Israeli accent. In English, he asked me: "Sex? For money?"

When I told him off in a very disgruntled Portuguese, he apologized repeatedly, saying: "I'm sorry, I'm sorry."

Immediately and inexplicably, my brain switched back to my native language. "You should be," I replied.

This experience was, unfortunately, one that I was mostly prepared for, given the scholarship on sex tourism as well as the experiences of Afro-Brazilian women I had met.[35] I knew that as a black woman standing on a street corner, although I was waiting at a bus stop, foreigners might perceive me as a prostitute. However, as a Christ-follower, it was startling to be approached by a man dressed like the Orthodox Jews I had come across in the United States who I assumed was religious as well. Intellectually, I knew that religious people have been participants in the sex trade for centuries, but, spiritually, it hurt more because of his attire.

Navigating Africanness

As a dark-skinned black woman sporting short dreadlocks, I could blend in as a Brazilian, even a Bahian (*baiana*[36]), as long as I did not speak. However, after one year of studying abroad and another as a Fulbright Scholar in Spain, my use of Iberian cognates, along with my non-Brazilian accent, often led Brazilians to read me as a native of Angola or Mozambique. On one occasion, I was even chastised for not expressing pride in the beautiful capital of Luanda in my supposed home country, Angola.

On the one hand, I was proud that I did not have a strong American accent and that I could pass off as a native speaker. On the other, as a second-generation immigrant with roots in the most populous country in Africa, my pride was stung when I was attached to other, much smaller countries. This mislabeling and "cultural misrecognition"[37] left me uncomfortable and would often result in me clarifying how my parents were immigrants from Nigeria to the United States.

Some Brazilians were excited by my Nigerian ancestry. These often included Brazilians who practiced African-derived religions like Candomblé and Umbanda, participated in cultural practices like black percussion groups (*blocos afros*) or the Afro-Brazilian martial art of *capoeira*, or those who were affiliated with the black movement. Apart from being black, we had a bond of recognition and valuing of African ancestry and ethnic identity that was refreshing.

However, often the first question that popped out of black Brazilians' mouths, upon learning that I was Nigerian, was "Are you Yoruba?" The first time, I was taken aback. Even in the United States, first-generation Nigerian immigrants draw strong ethnic lines between different tribes. The Biafran War, the civil war that raged in Nigeria from 1967 to 1970, took place in my parents' lifetime and they still recount stories about "what 'they' [meaning other tribes] did to us." As a second-generation Igbo-American, being automatically asked if I was Yoruba was always a little unsettling, given this troubled history and current reality of ethnic boundaries, even in the United States. My idea of what it meant to be Nigerian clearly did not line up with their reflection.

For the Brazilians whom I encountered, what it meant to be "Yoruba" had a very different meaning. Transatlantic slavery brought millions of Africans to Brazil between the sixteenth and nineteenth centuries. Of the many tribes who arrived, the most influential were the Yoruba, whose culture had the greatest influence across Latin America and the Caribbean. Their excitement at our potentially shared ethnic ancestry dissipated a little when I replied: "No, I'm Igbo."

The conversation usually ended there. If I was not Yoruba, people did not know what to do with me. There was a lack of knowledge about Nigeria and its major tribes (Yoruba, Igbo, Hausa) or that a civil war along ethnic lines tore the country apart decades ago. I am by no means an expert on Nigeria or Nigerian politics and many of my relatives have told me that I am "very Americanized." Yet, it stung, just a little, to be associated with a tribe toward which my family members have some hostility. The only US analogy might be the sentiments that older African Americans have toward white Americans, given what they had suffered at the time of Jim Crow. If ethnic identity truly waxes and wanes depending on the circumstances we find ourselves in, I never felt so Igbo as when I had those experiences in Brazil.

However, the lack of recognition went both ways. When I was in Rio, my friend Rodney explained to me that his origin was Bantu. I was familiar with the term Bantu as a category in the family of African languages and tribes or as a group affiliated to colonial Brazil.[38] However, I was not familiar with Bantu as a tribe or ethnic group that meant anything today. My ignorance was evident in my failure to recognize what his Bantu origin meant to him. It was only later that I learned about non-Yoruba Afro-Brazilian political and religious organizing during and long after slavery.[39] Along with the Yoruba, other African "nations" created their own religious institutions and mutual aid societies to purchase the freedom of enslaved persons and care for the sick and elderly. These nations influence the social organization of Afro-Brazilian communities even today, having embedded themselves in political struggles to help their communities.

I was not prepared to negotiate my identity as a Nigerian American while I was in Brazil. While some scholars have discussed issues of identity surrounding fieldwork in Africa,[40] the reverse situation of Africans doing fieldwork in the diaspora (outside of the United States) was not something I had anticipated as an issue.[41] My looking-glass African self was obscured by the different meanings that Africa conveyed for me and for Brazilians. This made it difficult to anticipate the ways that Brazilians understood me as a person from both the United States and Nigeria. Neither of us could incorporate the other's judgments into how we interacted with one another, which meant that these conversations that I had were often stilted and ended quickly.

Conclusion

The dictum that sociology has the ability to "make the familiar strange" is easily understood when conducting research in an unfamiliar society. Unfortunately, as nonwhite women, the "looking-glass self" that we are supposed to be aware of can be very foggy and have cracks in it across the different statuses that comprise ourselves. As a foreigner, my "looking-glass self" was unclear and my "double-consciousness" was not developed for Brazilian society. I was speaking another language and was a cultural outsider who did not always know how to read cultural (gendered) signals; as a result, others misread my intentions. The different social categories to which I belong intersected in new ways in Brazil, revealing how, for US black women and other women of color, our double-consciousness can become distorted when conducting research abroad.

Transitioning from fieldwork in Los Angeles to Rio de Janeiro, and Brazil more broadly, revealed a "looking-glass self" that I did not often recognize. No longer was I an angry black woman investigating black–white couples; in Brazil, I was supposedly pursuing intimacies with white men, whether romantically or, as one encounter suggested, as a profession. In addition, the US-based immigrant meanings that I gave to being Nigerian were complicated when they brushed up against Brazilian notions of Africanness. Furthermore, the African selves that Brazilians and I reflected back to one another were at times mutually unrecognizable.

My inability to read how others saw me led to some awkward encounters in the field. In addition, I was not always cognizant of changing my behavior in an attempt to control how others read me as a researcher. Even though I usually saw myself as a "sociologist in the field," I was not always self-conscious about being a woman, and specifically a black woman, in the field. While some black scholars have discussed issues of sexuality while conducting research in Brazil, the ways that gender intersects with race and immigrant background for researchers of various colors have not always been explicit.

I am sure that my citizenship as an American and as a university-based scholar living in upscale neighborhoods protected me from injury on more than one occasion. Nevertheless, my shared race and gender with women in Rio de Janeiro led me to have experiences similar to Brazilian *morenas* and, perhaps, other Africans (particularly Angolans and Mozambicans) in Brazil. It is my hope that future scholars will be more cognizant of the different ways that social categories can color our experiences in the field in unexpected, yet messy, ways.

Notes

1. When I use the term interracially, I refer to people partnering across phenotypic color, regardless of how they identify themselves. This was different from the research that I conducted looking specifically at black–white couples, just one type of interracial couple.
2. Clarke (2011), Feliciano et al. (2009), Qian and Lichter (2011).
3. Freyre [1933] 1986, Joseph (2015).
4. Contins (2005).
5. The names of respondents and informants are pseudonyms.
6. Kennedy (2003).
7. Fernandez (1996), Moutinho (2004), Chito Childs (2005b).
8. Americas Barometer (2010), Silva and Reis (2011).
9. Fanon 1952 [2008].
10. Harris-Lacewell (2004), Chito Childs (2005a).
11. *Loira gosta de negão.*
12. *O gringo adora uma mulata.* The term *mulata* can also refer to Afro-Brazilian women writ large.
13. Caldwell (2007), Collins (2004), Osuji (2013).
14. In Brazil, the word *gringa* or *gringo* applies to any foreigner, regardless of their color. For this reason, I, too, was considered a *gringa* when people knew that I was not Brazilian.

15. Goldstein found a similar preference of Afro-Brazilian women for white men, particularly foreigners, in romantic relationships.
16. For example, I was once called a *morenona*, an augmentative of *morena* by a white taxi driver.
17. Ribeiro and Silva (2009), Petruccelli (2001), Telles (2004).
18. *Preto* directly translates as the color black and is often used to refer to Afro-Brazilians with the darkest skin tones. The word *negro* often refers to the black race and may encompass people who identify both as *pardo* and *preto*. See Telles (2004) for further discussions of the terms *preto* versus *negro*.
19. Guimarães (2005).
20. Nogueira (1985).
21. Osuji (2014).
22. Barros (2003), Moutinho (2004).
23. One husband out of the 25 couples could not participate.
24. Telles (2002, 2004).
25. Cooley (1902).
26. Du Bois (1903).
27. Chito Childs (2005a).
28. Collins (2004).
29. Twine and Warren (2000).
30. Telles (2004), Sheriff (2001).
31. In terms of my research, this incident demonstrated that the driver did not consider his wife as a *negra*, despite her racially mixed ancestry. This was a theme that would emerge in my research on the "black" wives in these relationships in Rio, in which some remarked they were not always seen as "really *negra*."
32. *Morenona* is an augmentative of the word *morena*.
33. His initial response became a part of the way that I recruited respondents, looking for couples involving *brancos* and *negros*.
34. Nogueira (1985), Guimarães (2005).
35. Williams (2013).
36. *Baiana* is a term that is not only exclusive to people from the region of Bahia. It is sometimes also used to refer to people of dark skin and of African ancestry.
37. Fernandes (2011).
38. Butler (1998).
39. Pinho (2012).
40. Pierre (2013), Talton and Mills (2011).
41. However, see Dawson (2014) on the ways Africans negotiate Brazilian notions of Africanness in Salvador da Bahia.

PART III

Black Brazilians' Reflections in the United States:
Myth of a Racial Radical Paradise

CHAPTER 10

Living the African American Way of Life—Impressions and Disillusions of an Afro-Brazilian Woman in the United States

Daniela F. Gomes da Silva

Love at First Sound

Love at first sound. This is how I choose to describe how my journey as an Afro-Brazilian living abroad started. It probably started with a mix of sounds, such as the sound of the Christian hymns that I used to listen to growing up in a Baptist church, or with my participation in the church choir during which I used to dream about joining a choir like those on TV, or perhaps with my time listening to the Ray Charles songs that my father used to listen to at home. I cannot remember exactly where my contact with African American culture started, but I am absolutely sure that it was a sound that awakened in me a passion for a place that I did not know. I still remember the first time I heard some words in English; maybe it was not the first time I had heard them, but it was probably the first time I had paid attention to them. I was nine years old, growing up in a poor community in the city of São Paulo, Brazil. My family was hosting dinner for North American missionaries who were visiting our home church. When I heard those words, I fell in love with the language, and in future years it would take my imagination far away from my home and my reality. Learning English became my priority, and without financial resources, I self-taught myself by listening to music and translating songs from English to Portuguese with the help of an old dictionary. A few years later, still a young teenager, a new sound would make a difference in my life.

When I first heard the rhymes of a famous Brazilian rap group, my love for the language found a priceless partner. Through my connection with the hip-hop movement, I started hearing about African American heroes who made a difference in the world through their work in political and cultural fields.

Angela Davis points out that music produced by enslaved people in the United States was a "collective" production that "helped to construct community" (1999, 5). I daresay that the influence of the hip-hop movement as well as other African American music genres brought to my life this sense of unity (or to quote Queen Latifah, "U.N.I.T.Y."). It was as if from that moment I was part of something bigger that went beyond my comprehension of the world and completely changed my life. It was not only the first step to recognizing my blackness; I also consider that moment to be the beginning of my life as an activist against racism and the oppression of black people in my country and, eventually, in the world.

This feeling of becoming integrated into a community can be approached with the perspective offered by Stuart Hall, who says that one of the manners in which diasporic people can think about "cultural identity" is through the view of "one shared culture." According to the author:

> A sort of collective "one true self," hiding inside the many other, more superficial or artificially imposed "selves," which many people with a shared history and ancestry hold in common. Within the terms of this definition, our cultural identities reflect the common historical experiences and shared cultural codes which provide us, as reference and meaning, beneath the shifting divisions and vicissitudes of our actual history. (1994, 393)

In this sense, those other people with whom I share a past of slavery and oppression were an important part of my process of growth and identity formation, because in my mind, they were brothers and sisters who overcame barriers. In a country like Brazil, where even if there are more than 100 million people of African descent (which represents 51 percent of the population), the media gives little visibility to the black population, especially in terms of positive images, to help develop self-esteem. Yet, this is what I did see in African American productions, especially rap videos, which became my reference for positive images of blackness.[1] According to Donalson (2007), movies were an important tool to represent the hip-hop movement in the 1990s. As a child and a teenager in the 1990s, my life was also full of these images—one of my biggest dreams was to see a lowrider car with my own eyes and check if it would really jump while rap music was playing. Of course, this was not the only identification that was possible for me; the common history of racism and poverty also meant that the people in those movies looked like they were closer to my reality. Audre Lorde affirms that for "[b]lack women . . . poetry is not a luxury. It is a vital necessity of our existence" (1984, 37), and as it had both rhythm and poetry, rap music became a fundamental part of my life through which I found the reflection of my image, the healing for my pain, and the answers to my search.

If, as pointed out by Hall, identity can be perceived as something that is in constant "process" throughout "representation" (1994, 392), it is possible to say that over the years my identity was formed through the representation of African American culture and historical struggle as synonymous with blackness. My wish was to see that with my own eyes, to meet my brothers and sisters and learn from their experiences. I imagined talking with them about our commonalities and understanding how we, as Afro-Brazilians, could "get there" to the place where black people in Brazil would be much better off than they were here, as I used to think.

In my master's thesis, "*O Som da Diáspora—A Influência da Black Music Norte—Americana na Cena Black Paulistana*" (The Sound of the Diaspora—The Influence of American Black Music on the Black Scene of São Paulo), I discuss the way a group of Afro-Brazilians (including myself) had constructed their racial identities based on ideas of blackness that we had seen or observed in the United States. I examine the images from the United States that these Afro-Brazilians used to imagine what they believed was the reality for the African American population in the United States. I term this imagining about what it means to be an African American and the search for this lifestyle as the "African American way of life" or a "black version of the American dream" (Silva 2013a, 87).

My impressions and disillusions, as someone who ascribed to the idea of the African American way of life and had the opportunity to attest to the reality with my own eyes, is what I present in this chapter. In the first section, I give a brief history of how this image of the black United States was constructed abroad over the years. In the second section I discuss how my desire to know more about the African American reality led me to an experience in Atlanta, Georgia. In the third section, I describe how when I returned to the United States for a second time as a graduate student in Texas, my experiences were sharply different.

The African American Way of Life as Synonymous with Blackness

Throughout the twentieth century, one of the most widespread representations of American society around the world was the image of its black population. Through the struggle against racism, the cultural, social, and political perspective of the African American community crossed borders and showed the world what it supposedly meant to be black in the United States. These iconic images include those of Martin Luther King Jr. leading the March on Washington, of Angela Davis as a fighter during her participation in the Black Panther Party, as well as of Diana Ross and The Supremes. Most of the content that was spread reflected a revolutionary world. Through these illustrations, the world received an image of a united black population in the country and a definition of blackness associated with a more complex and broader notion of belonging. As I mentioned before, it was this sense of belonging that motivated me to learn more about African American culture and history at different moments in my life. However, I was not the only person who was touched by this image and its deep meaning.

Though the twentieth century was a time for an intense propagation of African American culture, it was commonly known that, most of the time, the treatment of Africans abroad did not reflect the hardship that they faced in their own country. Since the end of the nineteenth century, individuals such as Ida B. Wells-Barnett, W. E. B. Du Bois, and Frederick Douglass have described fascinating narratives of their experiences outside the United States. Douglass expressed how amazed he felt with the treatment he received during the time he spent abroad (Meyer 1984). In a similar way, Wells-Barnett (1997) describes her happiness when she realized that her denouncements on black lynching in the United States were well received during the time she was visiting and writing articles from England. The feelings these authors expressed were no exception to the descriptions, during the same period, of several African Americans who had the opportunity to leave America for Europe and find a place of relief from the oppression they faced in daily life.

The experience of receiving better treatment abroad is also depicted by Saidiya Hartman in her book *Lose Your Mother* (2007). While narrating her experiences in Ghana, the author tries to revive a memory that was lost as a consequence of the enslavement suffered by her ancestors. She realizes that, on African soil, Ghanaians perceive her as a different kind of black person, one who achieves things that are only part of the imaginings of that group. In her words, she realizes that "[t]he country that most of us had come running from was the one of which they dreamed. They would have traded places with us in the blink of an eye" (104). In other words, one could possibly say that being a person of African descent in the United States sounded much better abroad than it sounded for those who were born and raised in that country.

If, as explained by Nell Irvin Painter (2006, 5), it is possible to affirm that "being 'African' American is part of a New world identity," which was constructed in a long process of trying to recover a history and memory that were lost during the travel inside the slave ship, it is also possible to affirm that the African American community brought a new meaning to the understanding of blackness through its political visibility. In this sense, the second half of the twentieth century was a platform for new manifestations that served as an inspiration for people of African descent around the world.

As one of the main expressions of this new meaning of blackness, we can point to movements that were not only political expressions but also foundational in the cultural transformation of American society. An example of this type of movement can be found in the spread of the black power movement and its symbolic representation around the world in the early 1960s. Its message of black pride was able to reach different groups around the world and generate similar manifestations in different places of the globe (Painter 2006), including Brazil, where people from different cities found in the movement an echo of their own struggle. Furthermore, if, on the one hand, it is possible to affirm that the political face of the movement made it possible to communicate with different people around the world, then, on the other hand, it is also possible to affirm that the movement's cultural point of view was probably responsible for its propagation outside the United States. Painter elaborates:

> During the Black Power era of the late 1960s and early 1970s, masses of African Americans—workers, intellectuals, artists—looked inward. They broke from the prevailing American mind-set of seeing black people as a problem and found beauty and value in blackness. (292)

This main expression of black pride reached artists with an international visibility who incorporated the message into their work throughout the country and made it possible for the black world outside the United States to understand how this could be applied to their own reality. In the subsequent years, the image of the African American community would be transmitted to the world through the lens of the media (Birdwell 2009). Propagating a successful image of the African American population, the black community spread out a new concept of blackness, which I call the "African American way of life."

If one can say that the cultural perspective of the African American struggle was fundamental in promoting an image of a strong community outside the country, it is also feasible to affirm that music was the main expression of this culture. Thus, the message *en vogue* in the country reached many more people through the music. According to Samuel A. Floyd Jr. "African-American music has always mirrored a wide range of struggles and fulfillments . . . all these experiences reflect the gropings, failings, and inhibitions, the successes and pleasures, regressings and advancings of the black experience" (1995, 227).

As a carrier of such powerful meaning throughout the history of the black population in American territory, music in the second half of the twentieth century became, itself, the expression of black pride that was spread in the middle of the streets as a claim for justice and equality. It is also possible to affirm that, combined with a new way to portray the black population and the changes that were happening in the country, the advancement of new communication technologies led to an extremely important occasion to extend this feeling of black pride beyond the frontiers of the country and reach a global audience. In his book *The Black Atlantic* (1996), Paul Gilroy illustrates the experience of the South African leader, Nelson Mandela, who during his time on Robben Island found some release from his suffering "listening to Motown music," as a way to express how a diasporic flow works in a two-way street (96). Although the author did not state that Mandela formed his global understanding of blackness through this experience, it is possible here to imagine that the African American identity had spread throughout the world. In other words, it was not only I who was reached through this feeling of belonging. This diasporic perception of connecting through music was responsible for the idea of having different people around the world who not only completely understood the feeling presented by James Brown when he released "Say it Loud, I'm black and I'm proud," but also shared the same feeling of belonging.

In her book *The Bluest Eye*, through the description of some of her characters, Toni Morrison creates the following narrative to illustrate the low self-esteem that affects our people: "The master had said, 'You are ugly people.' They had looked about themselves and saw nothing to contradict the statement; saw in fact, support for it leaning at them from every billboard, every movie, every glance" (2007, 39). As a point of commonality for people in the diaspora, "the master's voice" was also able to destroy the self-esteem of the majority of Afro-Brazilian people, and I daresay that the contact with African American music, culture, and political movement was a fundamental tool to make our people stand up and declare our beauty and power and receive a new strength in the struggle. An example of the power of this representativeness can be observed in the way Afro hair is referred to in the country. The image of artists and activists with natural hair, styled with the Afro comb, was so associated with the image of the black power movement that, in Brazil, the hairstyle became known as "black power" (Silva 2013a, 148). Hearing someone use the expression that she or he "has a black power" became commonplace for those who, since the 1970s, decided to use the Afro hairstyle as a form of resistance against racial oppression (including myself at the present moment). As I mentioned before, my political awareness during this time was one of my main motivations for fulfilling my wish to see the United States with my own eyes.

"Georgia on My Mind": My Experience Living in a Chocolate City

My first contact with the hip-hop movement and black music from the United States became the first step in a search for knowledge about the black struggle in a country that guided me toward a life of activism from an early age into my adulthood, both personally and professionally. As a young adult with a degree in journalism, I started working at Zumbi dos Palmares College.[2] In this place, which, until the present, represents a real home for me, I began having a deeper connection with the African American community on different occasions when groups would visit the college to learn more about the conditions of Afro-Brazilians. Each exchange program that took place at Zumbi was a way to exchange information and learn more about the situation of the blacks in the United States. More than work, it was a pleasure for me to talk with the visitors and absorb the information that was available to me from anonymous and famous African Americans. With every passing day, my work increased my wish to see their reality with my own eyes.

Throughout these experiences, I was able to construct sincere friendships with brothers and sisters from the United States, and my wish to visit the country started becoming a goal. After a few years, the necessity of improving my English skills motivated me to invest in a trip abroad for a few months. Among several options of countries where English was the first language, I had no doubt that my choice of study location would be the United States. If Ray Charles's songs were part of my life from an early age, so, too, was the vision of a specific place in the United States. It was the birthplace of the civil rights movement and some of its most important leaders, or to quote Du Bois referred to as the "Gateway to the Land of the Sun" (1903 [1996], 77). When the time arrived, I surprised most people with my certainty about where I wanted to go for this exchange program; instead of opting for famous places that frequently received Brazilians, such as New York, Miami, or San Francisco, my choice was Atlanta, Georgia.

Visiting Atlanta for me was such a spiritual journey. As Mecca was the spiritual center for the Muslims and Canaan was the Promised Land for the Christians and Jewish people, Atlanta was the heart of the African American United States for me. I never knew why, but I was sure that I needed to go to that place. Hartman states the following about her trip to Ghana: "It's hard to explain what propels a quixotic mission, or why you miss people you don't even know" (2007, 17). And this was exactly the feeling that I had in my heart for most of my life, an anguish that pointed to my need to go to that place. I grew up with little information about my African ancestry. I did not learn about it at school or with my family; everything that I learned about my blackness, and mostly about African American history, was through my activism. It was that history that guided me to learn more about my heritage as a person of African descent, and I believe that it was one of the reasons that my search for the past did not guide me to the African continent but instead brought me to the United States to see the history of those who were my heroes in the struggle for freedom, those who inspired me and helped form my identity.

Hartman remarks that "the country in which you disembark is never the country of which you have dreamed" (2007, 33). As someone who dreamed of this trip to the United States for my whole life, I was sure that I would either love it, and my soul would find peace, or I would be totally disappointed and start to hate the place. But no matter what the outcome, I needed to face it. In 2008, before I left my country, I contacted all the people that I already knew in town so I would have some familiar faces to meet with. I packed and embarked on a journey that would mark me in different ways—not only by the importance of the trip itself, but also because it was the first time that I was leaving my country. Unlike Hartman's experience, my journey met my expectations. As soon as I arrived, the immigration officer was friendly with me, and having heard so many negative experiences, I took this as a positive signal which indicated that it would be an incredible trip.

Since I was a student but not formally in school, it was now time to do what I had come to do: see Atlanta with my own eyes, hear the noises, know more people, and visit places. And I did. My first surprise was realizing that, there, I was considered pretty. Morrison asserts that "[t]he death of self-esteem can occur quickly, easily in children, before their ego has 'legs' so to speak" (2007, x), and this was exactly what had happened with me, for several reasons (the most important being racism in my country). I was never perceived as a beautiful woman, and as a consequence, I never saw myself as someone who had beauty as an attribute. For this reason, to walk in the streets of the city and see people paying attention to me as if I were a beautiful person was something completely new. A second thing that caught my attention was the sense of black community. Black men and women who were strangers talked to me in the streets only because I was black; they did not even know me. The third thing that impressed me was that, there, I did not have to explain my blackness as I needed to do all the time in Brazil, because of my light skin.

Walking in the streets and seeing symbols associated with American black neighborhoods, like the barbershop and the beauty salon, or cars playing rap music loudly was like being in a movie scene. But if I must recall one simple experience of sound that was the most important to me, it would be my visit to a black church and the opportunity to listen to a choir as I have dreamed of my whole life. That was certainly my dreams coming true: the children singing with strong voices, the sisters with their hats, the spiritual manifestations, everything beckoned my attention.

Distinguished people guided me in my mission to visit places that were fundamental to the civil rights movement in the town. An activist named Joseph Beasley, who I knew through my work in Brazil, showed me some of the most important places for the civil rights movement. With his wisdom of 70 years, he taught me that we cannot be tired in the fight against racism, because we do not have the time for rest. Another friend took me to The King Center to visit the museum and see the places where King lived. To this day, I remember that as a very emotional experience. I cried almost every moment I spent in those places. One day I went to a restaurant where Beasley introduced me to an old man, Paschal, who was the owner. I learned that this was the first restaurant of the black community during segregation, and it was also where King used to hold some of his meetings. While in the restaurant, I started crying so intensely that I almost choked on my meal. Unfortunately, however, my tears were sometimes those of someone who felt hopeless after seeing the reality of scarcity, poverty, violence, and residential segregation, a part of the black community that represented the opposite of the situation that I had imagined before I left my country. I cried because, if the situation was still so bad in the United States, the country of the "African American dream," what hope could we as Afro-Brazilians have? The impact of this experience motivated me to try to understand the historical factors that facilitated the creation of this idealized image of the US and its propagation, not only in my country but also in other places.

However, such a rich experience would present more than simply a diaspora of pain. As I became more integrated in my new life, the meaning of the diaspora was also revealed to me through daily life experiences, such as with my introduction to traditional soul food. The food sounded so different that I had high expectations and yearned to try it, but when I finally tasted it, it was just more of the same, in the sense that it was so similar to the food from some regions of Brazil (as in the state of Minas Gerais, for example). This made me certain that our common African ancestry was present in those dishes. In the same way, my moments of leisure took me to some clubs in the city, where, instead of finding something really different from the clubs that I used to go to in São Paulo, I not only heard the same kind of music, but I also observed that most of the young people's behavior in those places—the way they talked, dressed, and behaved—was really similar to my reality with my friends in Brazil. After two months, I went back to my country (the tears were also part of my farewell), but the experiences that I had had and the strong sense of the diaspora I had felt went back with me and resulted in several changes in my life. If my journey started with sounds, it is certainly a particular sound, "just an old sweet song" that "keeps Georgia on my mind" and will not let me forget such an amazing experience (Charles n.d.).

Yes, I'm an Afro-Brazilian Woman, but What Does That Mean?

After my return to Brazil, my professional life followed unexpected directions. In a quest to improve my résumé and upgrade my career as a journalist, I enrolled in a graduate program with courses that led me to write about the impressions that were left on me regarding the commonalities among Afro-Brazilians and African Americans, especially the influence of American black music on young people like myself.[3] The project deepened my research on American black music and identity in my city. As I started my course, the possibility of finally combining my passions involving hip-hop, diaspora, black identity, and other subjects into one research project really fascinated me. However, I was sometimes advised that it would be necessary to separate my activism from my life as a scholar. At that moment, it sounded like self-mutilation. It was impossible to separate one thing from another when both were important parts of my being.[4]

I was still wondering how to reconcile these two things when I received an e-mail from some black activist women regarding an extension course in Theoretical and Political Questions of the Diaspora, which was a partnership between a Brazilian university, a nonprofit organization for black women named Criola, and the University of Texas (UT). The possibility of studying the diaspora was so amazing that, although the course was in Rio de Janeiro, I did not see any obstacle that could stop me from taking it. In that course, I had access to a syllabus comprised mainly of black authors whom I had never heard of before and probably would not have heard about were it not for this opportunity. After interacting with professors and students who were part of the program, I learned about the importance of activist research and saw that it was the answer to combining my activism and my new life in academia. This concept was so rich and so new that I decided that I would pursue it. Some months later, the diaspora came my way again when I was selected to participate in the first United Nations Fellowship Programme for People of African Descent in Geneva, Switzerland. During that course, with nine other brothers and sisters from different countries, I understood that my struggle against racism could not be limited to my own country, and that it was important to create a space for dialogue between black people around the world because our fight should be a collective fight. The diaspora was not an imaginary thing anymore; it was already a fact in my life and each day the desire of learning more about it was growing in me. Two years later I was accepted at UT, Austin, and so I finished my master's degree in Brazil and was ready to go back and start a new journey in the United States.

With the memories of my trip to Atlanta, I started to prepare myself to move back to the country, excited with expectations of the new course and my new life. In his book, *Imagined Communities* (2006), Benedict Anderson presents an analysis of "nationalism" in a framework of cultural exchange, observing the idea of a "political community" that would hover over a field of ideas. In his words: "It is imagined because the members of even the smallest nation will never know most of their fellow-members, meet them, or even hear of them, yet in the minds of each lives the image of their communion" (6). It was this imaginary connection that permeated my mind while I was preparing myself to face the challenges of my return.

However, my arrival in Austin, Texas, soon revealed that this trip would be totally different from my expectations. A gentrified town, which, according to Tang and Ren, is "the only . . . fastest growing major city in the country . . . that suffered a net loss in its African American population" (2014, 1), Austin did not offer me the same sense of black community that I had in Atlanta. With around 60,000 black people, this was not a "chocolate city," and I was not surrounded by friendly black faces. In my first week in the town, people touched my hair in the streets to see if it was real, and some even made absurd comments saying they had tried to wear their hair like that to a Halloween party once. These gestures reminded me of the stories I had heard during my childhood in Brazil of some dark-skinned black people who were touched by white people to see if their hands would become dark. My contact with black people was limited to my program at UT, and I became anxious and frustrated. Feeling that my blackness spoke louder than my nationality, I also did not feel motivated to pursue my relations with other Brazilians in the city or with the Latino community. Yet, within that chaos, the material I was learning at school was priceless and was an opportunity to focus on other things than my need to create a community. I had to face this adaptation process during my first semester, until I started to find spaces where I felt welcome, such as the UT Black Graduate Student Association and a Baptist church in the gentrified East Austin.

However, the absence of a black community was not the only disillusion that I suffered in this new experience in the United States. More mature and less impressed with life in the United States (although I was still in love with the country), I started to perceive that, as an Afro-Brazilian woman, I received a different type of attention from black men in the country. Of course, when I was in Atlanta, friends advised me that men were more inclined to flirt with Brazilian women, but I never took them seriously. For the first time, I started to realize that, with a Brazilian woman, especially if she was black, flirting had a different connotation. When the question "Where are you from?" entered the conversation and the answer was "I'm from Brazil," the person flirting became more sexual than charming and usually responded with a large smile and the expression "Uhn, Brazil." I had always heard rumors regarding the association of Brazilian women with prostitution abroad, but because my relationships with foreign people, especially African Americans, were always positive and friendly, I thought this was not the case for all women, and, moreover, I thought the association would be made only by white men.

Although Erica Lorraine Williams affirms that the concept of "'controlling images' cannot necessarily be applied to Brazilian reality" (2013, 114), in my understanding in the United States Afro-Brazilian women were perceived as a tropical version of the Jezebel, which according to Patricia Hill Collins is "a woman whose sexual appetites are at best inappropriate and, at worst, insatiable" (2000, 83). Afro-Brazilian women are not seen as respectable women. When you say you are Brazilian, you are not on the list of reputable ladies anymore and you are not worthy of serious attention; as a "whore," you are a second-class citizen. This mix of xenophobia and sexism provides space for experiences like the one lived by Williams (2013), who describes how she was treated as "quick and easy sex" just because she was a black woman in Brazil and was perceived as a Brazilian (she did not say she was American). In some conversations, I heard comments such as "I need to go to Brazil before I get married, because of the women there," and other stereotypical remarks. These statements made me sad not only because it excluded the possibility of a future relationship (this was not my goal in the country), but mainly because when African Americans said these things, the notion of community I was carrying with me—the perception that we were one black people in the diaspora—began to crumble. These feelings only got worse when, working as a translator on a trip to São Paulo, I witnessed some cases of sexual harassment which greatly affected me. I was shocked by these instances, and could not understand why things like this were happening, why black men and women in the United States could not see that we were facing the same battles and that we, as Afro-Brazilian women, were victims of the same "controlling images" that affected African American women (Collins 2000). My only consolation was the fact that I had so many African American male and female friends in the United States that I knew that these incidents were not reflections of the entire black population, and I knew that, like my friends, other black people there were more open and free from this prejudice.

Touched by my anguish, an African American friend shared with me what she thought might be an answer to my questions. Sending me a link to the documentary *Frustrated: Black American Men in Brazil* (Greeze 2011), she explained how sexual tourism in Brazil also involved black people in the United States and how it was reinforcing the stereotypes of Afro-Brazilian women as whores in opposition to American women as reputable. Watching the documentary, I realized that most of the women interviewed in it were really involved in prostitution as a form of work. There were also stereotypes attributed to Brazilian women by the interviewees, as if we were less educated, or less demanding. The comments on YouTube and the articles about the video also followed along the same lines of not only treating Brazilian women as inferior, but also criticizing African American women, blaming them for those men's decisions, and creating a rivalry between us as black women in the African diaspora.

As a way of showing my indignation and sharing my opinion about everything that was happening, I wrote an article titled "About Being an Afro-Brazilian Woman Living Abroad," and posted it on my blog (Gomes 2013). At the same time that I was feeling lonely and hopeless because of the stereotypes and prejudice that I became aware of, I forced myself to remember that these attitudes were not held by all African Americans, that we were still one people, and that I could write to people as a voice asking for help. I did not want to give up on my imaginary diasporic community. I was really impressed by the number of answers that I received to my post. The comments came from different sources: from several Afro-Brazilian women sharing exactly the same feeling who decided to tell me their experiences and reactions, from African American women indignant about the situation and offering us their support, and also from African American men, grateful for my courage to talk about the subject in an open way and reject those stereotypes. As the post was republished by the blog *Black Women of Brazil* (www.blackwomenofbrazil.com), it also received several comments, most of them positive, encouraging me to keep up the struggle to change this image. This support was reassuring and brought back hope, and instead of losing my faith in the power of the diaspora, I understood that I was able to reach my brothers and sisters when I created a frank dialogue with them, without illusions or fantasies, but with the sweet flavor of reality, respecting the particularities of our cultures.

Conclusion

As highlighted in this chapter, my trip to the United States was motivated by "the African American way of life," by a black version of the American dream, and by a search for a mythical place, which I found to be somewhat different from what I was expecting. But if, on the one hand, my experiences did not guide me to the destination that I expected, on the other hand, it guided me to something bigger, showing me the power of the diaspora, how similar we are, and how strong we can be when we finally understand it. If, before my first trip to the United States, I was living under the notion that we were one people and that it was enough to connect us, after my new experiences I came closer to thinking in another way about the diaspora, which is a form that "recognizes that, as well as the many points of similarity, there are also critical points of deep and significant difference which constitute what 'we really are'" (Hall 1994, 394). I had started with the perspective that I need to highlight our differences and be able to create bridges that will facilitate a dialogue that will lead us in a common battle against racism and oppression. In my journey I found many more allies than those who discriminated against me, and this is what I carry with me—each hand that extended a gesture of friendship. Considering that I still have a long way to go in this new journey in the United States, I do not think it is possible to draw a conclusion about it. So, what I can say is that my relationship with the diaspora is under construction and it adds a brick every day, every time I ask for a diasporic connection and receive an answer from brothers and sisters throughout the country. In this sense, each step is fundamental to building this path, including the process of writing this text.

Notes

1. Though rap videos later began to focus almost exclusively on the hypersexualization of black women's bodies, the earlier phases of rap focused more on presenting the racial realities of African Americans. It is this aspect that reflected my personal experiences.
2. This is the only black college in Brazil, the creation of which was inspired by the history of the Historically Black Colleges and Universities in the United States.
3. The courses were graduate courses in Media, Information, and Culture, a specialization promoted by the Centro de Estudos Latino Americanos em Cultura e Comunicação, of the University of São Paulo and a master's in Cultural Studies also by the University of São Paulo.
4. One of the causes for this necessity is the absence of black studies in Brazilian education as a degree.

CHAPTER 11

Increasing Resilience to Face Diversity: Race in Academic and Social Environments from Salvador to Los Angeles

Lúcio Oliveira

Act One

It was two o'clock in the afternoon on a beautiful sunny Friday in 2011. After spending office hours with my adviser, I crossed the university campus on my way to the bus stop. There was a UCLA—University of California, Los Angeles—Police Department car parked ahead, with a police officer standing outside. As I passed by, he stopped me and asked if I was a student. For a moment I thought that I was getting an international stop and frisk by the Brazilian police. Between ironically laughing to myself and trying hard to keep the anger from showing in my eyes, I looked at the police officer, and instead of answering, I asked him: "What's going on around here?" He told me that "they had a situation on campus," and he was just checking (on me). We remained silent for a few seconds, looking at each other. I pretended that I did not understand his explanation for stopping me. The question "And why have you chosen to check on me instead of other (white) students around here?" was screaming in my head. My strategic silence and the fact that we were on UCLA campus probably created tension for him. He started showing embarrassment and hastily said "Never mind!" and drove away in the police car.

Everything happened very fast. In less than one minute my world had turned upside down, and I went from being an international student who got admission to a prestigious university to a suspect for robbery on the university campus. The police officer was Mexican American and, for me, it was not difficult to recognize that he was operating with the mind-set that black bodies are a threat. In the past, only police officers in Brazil had stopped me. So, being at a university like UCLA, which is surrounded by extremely wealthy and safe neighborhoods such as Westwood and gated condominiums like Bel Air, it might not be very safe for blacks. After checking on my progress in the program with my adviser, I experienced hell in a situation that was not surprising at all, although it was the type of situation that happens when one least expects it.

Act Two

On May 4, 2015, after reading a newspaper article by the renowned historian Robin Kelley (2015) on the recent protests in Baltimore, I read a Twitter hashtag that grabbed my attention: "Rest in Peace, since you couldn't be black in peace." I cannot remember what led me to that page, but I immediately started thinking about episodes of racism throughout my life, like instances of black people being killed by the police or security guards in the United States (Trayvon Martin: Florida, 2012; Rekia Boyd: Chicago, 2012; Michael Brown: Missouri, 2014; Walter L. Scott: South Carolina, 2015; Freddie Gray: Baltimore, 2015)—people whose tragic death had captured the attention of the American and international media. The Black Lives Matter slogan "This Is Not a Moment, but a *Movement*" really makes sense in light of these deaths.

One could try to comfort me and minimize the ravages of racism by saying that writing this chapter proves I won those battles against racism since I did not die in the interaction with the police officer. Those same people would probably add that I should focus on my academic achievements instead of feeling resentment for the past. I totally agree about nurturing positive feelings instead of negative ones, but that "past" is not the past, unfortunately. While I write these lines my motivation and what makes me recall memories is not only my personal story, but is also connected with all the recent episodes of police violence and brutality against black people and communities in the United States. Being aware of that makes me (once again) realize that there are institutional impediments that make "being black in peace" difficult and that the struggle for staying alive has not ended for black people in the United States or on Brazilian soil.

Understanding Race Relations through African American Music

I started the PhD program in the Department of Political Science at UCLA, in 2010, and in this new city in which I would live for five years, race relations also played a major role in history, social dynamics, and everyday life.

I had the initial challenge of making a transition to different cultural norms and a very distinct, and much more competitive, academic environment. But previous chapters about my intellectual trajectory help to understand how my current journey started so that, the more I live, the more I see their remarkable influences. My contribution here revolves mainly around a discussion of my introduction to diasporic consciousness through African American music, my participation in diasporic collaborations with African American scholars and graduate students, and my experiences in Brazil and the United States with the policing and surveillance of black bodies. With regard to the latter, I discuss how the policing of black bodies involves both racial prejudice and stereotypical perceptions of threat.

I was first introduced to African American culture through its remarkable musical expressions of the twentieth century, such as rock 'n' roll, blues, jazz, soul, funk, and R&B. I grew up listening to these African American music creations. The powerful vibes of electric and acoustic guitars coming from the speakers were the most beautiful sound I had ever heard. They gave me a childhood that was full of joy and strongly immersed in music. Thanks to this passion for music, at the age of 14 I taught myself how to play the acoustic and electric guitar. I associate that time period with such strong emotions and experiences that would influence my life forever. Back then, there was space only for the music, the guitars, and me. For some strange reason, it never crossed my mind that taking guitar lessons could be helpful. I ended up learning how to play the guitar by ear, figuring out songs' chords and melodies. I believe that my self-confidence in teaching myself to play the guitar was heavily influenced by African American music icons whom I admired, such as Chuck Berry, Ray Charles, Michael Jackson, James Brown, and Jimi Hendrix. They had attitude, and so did I. Well, at least I thought so at the time. Alongside this music, I also listened to Afro-Brazilian music, among other styles of the so-called Brazilian popular music genre. The youngest of four brothers, every Saturday I had to fight for the turntable because all of us wanted to be the DJ. Being exposed very early to African American music, I developed a positive association with African American culture and people.

To clarify a bit more about how the universe of race relations was unveiled to me, it is relevant to mention that until I was 18 years old, I lived in a small city in the countryside of Bahia, not too far from the beautiful coast of northeastern Brazil. It is about 60 miles from Salvador, the capital and biggest city of the state of Bahia. Despite not being too distant from Salvador, the city where I grew up was, and still is, one that has not been receptive to Afro-Brazilian history and culture; nor would I characterize the city as being particularly liberal. In the absence of these more racially affirming messages, my understandings of race were mainly shaped by daily processes of socialization that most Brazilian black children and adolescents go through: the denial of the existence of racism, silence about existing racial inequalities, and contradictory information about the racial boundaries between blacks and whites in Brazil's racial democracy[1] (Peña et al. 2004; Hordge-Freeman 2015b). My experiences with racism started in my early childhood and were mainly from the elementary school environment, social environment, daily life in the neighborhood, and interpersonal relationships. Among these, school is one of the primary environments in which children learn how to deal with questions posed by their social reality—implicitly and explicitly, pleasant or unpleasant. If my teachers remained silent and denied the existence of racism, it would affect me, a child who defied colorblindness explanations and attitudes. Such situations would lead to overwhelming confusion for me, since there was no room for discussion or any person with whom I could talk about these experiences. I had to keep to myself the anguish and doubts about how I was being treated and insulted based on my hair, skin color, and other physical traits.

When episodes of racism are not handled properly and continue to happen throughout one's childhood, there is a great risk that these episodes will shape one's personality. Some children may develop peculiar behavior due to an inability to respond to a continuous threat coming from the outside (Oliveira 2007). Children do not have—and are not supposed to have—the ability to protect themselves against racism and the conflicting feelings of inadequacy that come from it. Should teachers at schools not be prepared to deal properly with such situations and protect children, instead? Several scholars have already pointed out the harmful consequences of nurturing silence about racism in spaces of education (Tatum 1997; Cavalleiro 2000). The school environment was not one in which I felt really welcome, emotionally comfortable, or safe to express myself. As a result, I became an extremely shy and introverted adolescent. This is not necessarily bad; but if what leads a person to behave in this way is related to racial prejudice and discrimination, then these effects can jeopardize their educational trajectory and other kinds of achievement throughout life.

In Brazil, powerful agents such as the media and schools still indoctrinate the majority of the population to believe in a deracialized society. Nevertheless, daily experiences challenge us to think about the implications of our racial belonging and the racial association society attributes to us (Oliveira 2014). The impact of being black in multiracial societies like Brazil and the United States, specifically in school, is huge. Afro-Brazilian children's school performance can be affected in specific ways related to race and skin color. White children are not harmed by their racialization, as it has been shown that racial factors affect school proficiency independently of other factors, such as family income (Castro and Abramovay 2006). In this scenario, black children may be in a highly unfavorable situation, as, aside from the challenging tasks and the effort already required of all children, black children still have the extra burden of dealing with the acceptance or rejection of their racial identity (Oliveira 2007).

African American music provided a space for me to explore my inner dialogue that was not being addressed elsewhere. Music would serve as material for interpreting and understanding both the nature of my blackness and reasons for antiracist black movements that took place in US streets during the 1960s and part of the 1970s. I can recall that each album's front and back covers represented a new universe of information to dig into: iconography (the emblematic Black Panthers' salutation with clenched fists), hairstyles, clothes, bands, musical instruments, titles of songs, vocal expressions of chorus lines, the meanings of lyrics and liner notes. Both the visual and auditory aspects of the music functioned as gateways to a greater understanding of my racial experiences in Brazil.

While I grew up listening to a lot of African American music, I was able to develop some basic English language skills on my own. Through very raw and "choppy" lyric translations, the messages of protest against racial and socioeconomic inequalities in the United States still managed to come through and they helped me to understand that African Americans wrote those lyrics while singing and dancing beautifully in order to build a positive racial identity and defy negative stereotypes (Neal 1999). Music was the seed of my diasporic consciousness that would shape my intellectual trajectory. Later, when I began to research race relations in the United States for my master's thesis, several books that I read on African American social movements would confirm that what I sensed through music was a connection to both shared feelings and shared experiences.

On my side of the Atlantic, diasporic connections were also taking place around me. I would pay attention to the physical traits of and similarities between African Americans and Afro-Brazilians. In fact, some neighborhoods of Salvador had received the funk/soul message and combined it with their own material for fighting against racism in Brazil. This is the case of one of the first and greatest successful songs by Ilê Aiyê: *Mundo Negro*. The lyrics speak of the particularities of this collective of Afro-Brazilians, including their attitudes against racism, the appreciation and beauty of their skin color and hair, and, most of all, the pride of being black (Pinho 2010). Very consistent with the climate of protest against the military dictatorship that ruled the country during the 1970s, which dissolved black organizations and put them on surveillance, Ilê Aiyê used the expression *black pau*, an English–Portuguese mixed translation of "Black Power," to demonstrate their identification with the struggle for racial justice sustained by the black power movement (Covin 2006). An additional explanation for using the expression *black pau* is the fact that the natural Afro hairstyle inspired by the US black power movement in Brazil became known as "Black Pau" (Essinger 2005), symbolizing not only aesthetic, but also body politics (Hanchard 1994; Iton 2008).

Indeed, during one of my childhood summer vacations in Salvador, I was visiting a place adjacent to the historic black neighborhood of *Liberdade*, which is also the home of Ilê Aiyê, where I noticed black people wearing high-heeled shoes and bell-bottomed pants, and with big rounded natural hair. It was amazing, as though Funkadelic's mother ship had landed in Bahia. All this made me see concretely that African American funk and soul artists had a message for black people in other multiracial societies around the world (Vincent 1996). The power of diasporic elements to carry messages and political symbolism abroad—a transnational communication—was fantastic. The power of this message was the reason a cultural movement started in Rio de Janeiro, known as Black Rio, which then spread out across Brazil, with special appeal and political importance for Afro-Brazilians (Alberto 2009).

Roots of Diasporic Consciousness

About ten years had passed, and except for black music, I grew up surrounded by silence about racial issues within my family, among friends, in school, and around my neighborhood. By the time I went to college, my questions and confusion about race relations had increased. The answers did not come and the existence of racism was denied or silenced by professors. The faculty of the Department of Psychology at Universidade Federal da Bahia (UFBA) was predominantly colorblind in that they believed that race or color had no influence on a person's life or experiences. Although we were in Salvador, a city with a majority of black population as well as an important place for understanding colonial history and the slavery legacy in Brazil, racial issues were not part of any course or syllabus. Also, the composition of the body of students in the department was predominantly white and middle-class, just like in the other more privileged and competitive departments such as the School of Medicine, School of Law, and School of Sciences at UFBA. In college, because of these racial silences, I was not really aware of white hegemony, nor could I explain existing patterns of societal and institutional racism. I also did not have a sense of how they could negatively impact several areas of my life, including my educational achievements and employment opportunities (Telles 2004). Reality always reminds us of the polymorphic nature of racism. The closer I was to graduation, the more I realized that college was just another step in learning to deal with institutional racism.[2]

One important personal milestone in standing up against institutional racism occurred after I got my second job. I had just been promoted to a higher position after about two years in the institution, and during a staff meeting, the supervisor openly complimented me for reaching goals and accomplishing tasks. Nevertheless, right afterward she said she was unsure about my growing curly hair and thought it might cause trouble in dealing with clients. What she said could be transcribed as follows: "Since you [are black and] have curly hair, just being a good professional is not enough." For her, short curly hair was probably as important as maintaining and improving my professional performance, if not more so. The perverse irony is the fact that Salvador is a city with more than 65 percent of black people, according to the 2010 census. Roughly applying the above rule means that because of these ideas about "boa aparência" (good appearance), getting a job in Salvador is, by definition, already an unequal competition between blacks and whites (Damasceno 2000).

My experiences at the workplace in Brazil and when I was stopped by the police at UCLA campus—mentioned in the introduction—both contain elements of what has been described by the racial threat theory as the black body being perceived as a threat. In a hegemonic white race context, the black body (and other nonwhite bodies) surfaces as a symbolic threat, as being essentially linked to "deviant" behavior (Blalock 1967).

It takes form in the social context through the reproduction of similar mechanisms of social construction and fabricated conceptions that nurture racism, such as normative discourses heavily structured through an idealized and hegemonic racial pattern of whiteness, which projects the black body as undesirable (Reiter 2010; Hordge-Freeman 2015b). Ware (2001, 2004) and Frankenberg (1997) write about the constructedness of whiteness, highlighting that it does not exist independently from a relationship with blackness. This "whiteness" was created—and has been recreated—based on a relationship not with a historically black, African, or African American culture, but with a "black culture" invented based on the repressions, projections, desires, and fantasies of nonblacks (Rachleff 2004, 100). The black body will never fit into the categories of whiteness, the very construction of which is changing over time. Race as a social construct is transformable and malleable. Anytime the white hegemonic racial pattern is challenged, either physically or representationally, by the diversity of race and ethnicities within our society, efforts to neutralize this challenge will be established to protect that hegemony. Whiteness is a one-way concept that allows otherness only to reify and value itself—the response to threats to hegemony and claims of human equality is fear and oppression to eliminate the threat. By definition, black bodies and black people, constructed as the ultimate "other," are portrayed as inherently threatening.

The episode of racism that I experienced in the workplace caused me enormous emotional distress. It triggered many negative feelings inside me that lasted for days. That same year, I applied for a master of psychology degree at UFBA with the goal of doing something more meaningful, not only for myself, but to help to expose and eliminate racism through my professional service. Ultimately, that endeavor would propel me toward an academic career in the field of social sciences involving research on racial and social inequalities.

Introduction to US African American Scholars and Students

During the master's program, I met a group of scholars and graduate students who would boost my academic trajectory over some years. They were members of an international exchange program called Race and Democracy in Americas: Brazil and the United States.[3] I was invited to take part in a 15-day workshop with them, through which I could acquire significant knowledge about methodology in research on race in the social sciences. The scholars were from different areas of the social sciences: anthropology, sociology, political science, economy, and education. I was the only student from psychology. They all offered their expertise through lectures, presentations, and discussions. The workshop was also a unique opportunity to be introduced to several people who would later support my intellectual trajectory in several ways. Among those people were the following individuals: Luiza Bairros, from Federal University of Bahia, and minister of the National Secretariat of Policies for Promoting Racial Equality from 2010 to 2012; my future adviser at UCLA, Mark Q. Sawyer, professor in the Department of Political Science; Dianne Pinderhughes, professor in the Department of Africana Studies and the Department of Political Science, and former president of the American Political Science Association (APSA) from 2009 to 2011; Silvio Humberto, professor in the Department of Economics at State University of Feira de Santana, founding member of the Steve Biko Cultural Institute of Salvador-Bahia, and elected official for the city council of Salvador in 2012; Ollie Johnson, associate professor in the Department of Political Science at Wayne State University; Edna Araújo, professor in the Department of Public Health at State University of Feira de Santana; Vera Benedito from Federal University of Bahia; Dyane Brito Reis, professor in the Department of Education at Federal University of Recôncavo Baiano; Raquel Souza, PhD candidate at University of Texas, Austin; and Tonya Williams, Assistant Professor of Political Science at Johnson C. Smith University.

I engaged in readings, presentations, and discussions of state-of-the-art methodologies in researching racial issues. Being among prominent scholars and students meant that the 15-day program functioned like an intensive and informative ongoing Q&A session, which catalyzed my research project and expanded my understanding of how a scientific approach could be used to address racial issues. Additionally, my participation in the group would provide me with the foundation to reevaluate my conceptualizations of race in Brazil and the United States, something I would not have been able to access in the Department of Psychology at UFBA. In this sense, my early relationship with African American music helped build the connection. Due to my knowledge of music and the civil rights movement, I was able to establish a prompt and consistent dialogue with the US folks. Moreover, I became part of that network and asked for guidance in advancing my project for my master's thesis. Through such mentorship I was able to keep in touch with these colleagues throughout the subsequent years. When I applied for PhD programs in the United States, these networks really supported me and helped me to cope with some of the stress of being an international student in a PhD program. They reassured me that the program was a challenge, but not an impossible task to accomplish.

On my first day at UCLA I attended the graduate reception. It took place in a huge auditorium with hundreds of students. I sat down near some black students who seemed to be getting along very well. During a break, one of the black students sitting near me turned and said: "Hey! How is it going? Did you attend UCLA for college?" I told him I was Brazilian and had just gotten off the plane about five hours ago. He looked at me totally surprised and invited me to join their group for the day. I felt comfortable and secure making contact and being with them rather than with the white students because they seemed to understand how detached I was feeling in my first days in that academic environment. We went to a meeting of the Association of Students of Color, and most people there were African Americans, Mexican Americans, and Chinese. I was the only Brazilian at the meeting. Seeing people of color united among multiple ethnicities was very different from what I was used to in Brazil.

Between 2010 and 2014, there were totally five Brazilians in my department. There, we were all Brazilians with variations in skin color. Although differences existed among us in terms of socioeconomic background and opinions about racial issues in Brazil, we discussed our perceptions and experiences involving racial issues back home. The city I come from in Brazil is one where the African culture brought by enslaved Africans during colonial times is evident in several spheres of daily life, social practices, and traditions. Back in the United States, we were all categorized as *Latinos*. Being familiar with critical whiteness studies, this situation soon made me realize that, as Brazilians living in the United States, we had to (re)negotiate our racial classification (Joseph 2014). So, many of the Brazilians who were white in Brazil would not enjoy the same privileges of whiteness in the United States that they had in Brazil.

When I was invited by the editors of this book to write a contributing chapter, I had several themes that I could potentially choose from. After some deliberation, I came to the conclusion that all of them tended to revolve around a predominant issue: being a black person living through the academic and social environments of multiracial societies[4] (Omi and Winant 1994) with an ongoing legacy of slavery. This explains my choice to explore my experience as an Afro-Brazilian graduate student attending a PhD program in the United States. Regarding racial issues, I never had any fantasy that radical differences between Brazil and the United States would ever exist. Of course, these countries have had different trajectories regarding how racial inequalities took place, but they still maintain a history of racism and the legacy of slavery.

In 2010, the Brazilian government was being celebrated in the United States and on a global level due to the increased profile of Brazil as an international power whose economic and political successes were competing on more equal terms with developed nations. In various situations, I witnessed North Americans expressing optimism for the Brazilian government's leadership and confidence that it would continue to lead the country toward social change in ways that would reduce the economic gap among social classes. But these conversations left me feeling ambivalent—I was willing to engage in such conversation but, at the same time, my optimism was very moderate in celebrating how well Brazil was actually doing. While I mostly heard about the government's socioeconomic programs, nobody voiced opinions about what Brazil's contemporary economic and political rise meant for persistent racial disparities or how public policies were impacting racial inequality. Similarly, very few people had information or discussed federal government initiatives that promoted affirmative actions in public universities.

In the same period, after two years in office, Barack Obama's "honeymoon" with conservative America was over. (Some would argue it was over after his second day rather than his second year.) Critics from both sides were critical of his policies and their views were disseminated with the aid of the mass media. Despite these critiques (many justified and others not) and the fact that the black population was strongly and adversely impacted by the 2008 financial crisis, the majority of African Americans whom I met expressed happiness over the election of Barack Obama, but were not greatly satisfied with the status of the country. I noticed that they were separating his symbolic importance (which was good) from the actual effectiveness of his administration. They were hesitant to show their support for what the administration had achieved, just as I was being careful and somewhat contradictory in my views about the Brazilian government's initiatives in diminishing the socioeconomic gap. In fact, as graduate students with knowledge about the pervasiveness of racism in the Americas, we were doubtful about what Obama's presidency would mean given the country's historical social inequalities rooted in race.

What I have shared here are the experiences, interactions, and insights I have gained through the paths I have crossed on my journey during this period of my life. The willingness that many people, including the authors referenced here, have shown in sharing their work has inspired me and motivated me to write this chapter. I hope my modest efforts to communicate my ideas and share my experiences might help others in their own trajectories.

Notes

1. The theory known as racial democracy (also referred to as Iberian exceptionalism) argues that, relative to the United States, the nations of Latin America are largely free from the ferocious racial prejudice that has characterized race relations in the United States for most of the nineteenth and twentieth centuries. Racial democracy theorists base these conclusions on the fact that, compared to North America, the nations of Latin America experienced a marked absence of postmanumission institutionalized racism (e.g., segregation and Jim Crow laws), a general absence of race-based group violence (e.g., lynching and other race-based hate crimes), racial protest, and a strikingly high rate of miscegenation. It is argued that, to the extent that racial inequality is still discernible in Latin American societies, this inequality is almost exclusively due to the residual effects of racially contingent resource allocation in the past and not due to the effects of ongoing racial prejudice (Peña et al. 2004, 749–762).
2. Institutional racism takes place when institutions and organizations fail to provide professional and appropriate service to people because of their skin color, culture, and racial or ethnic origin. It manifests itself in norms, behaviors, and practices adopted in the workplace, which are the result of ignorance, colorblindness, prejudice, or racist stereotypes. In any case, institutional racism always puts people of discriminated racial or ethnic groups at a disadvantage in accessing benefits generated by the state and other institutions and organizations (Instituto AMMA Psiquee Negritude, 2007).
3. Among the central objectives of the project were "(1) to foster collaborative, cross-national research projects on race and politics in the United States and Brazil, focusing on African-descended populations, involving NCOBPS members and Afro-Brazilian social scientists; (2) to encourage junior Afro-Brazilian scholars to study political science; (3) to highlight for Brazilian political and higher educational leaders the importance of increasing the number of Afro-Brazilians admitted to colleges and universities, as well as to advanced degree programs" (nbcops.org).
4. Omi and Winant (1994) define "racial formation" as "the process by which social, economic and political forces determine the content and importance of racial categories, and by which they are in turn shaped by racial meanings" (61).

CHAPTER 12

Far Beyond "Fresh Prince of Bel-Air": Impressions from an Afro-Latina Filmmaker and Activist in Philadelphia

Gabriela Watson Aurazo

In this chapter, I analyze my experience as an Afro-Latina activist and graduate film student, during my two years of residence in Philadelphia in the United States. I discuss my reflections about my expectations, experiences, and moments of connection with the African American and Afro-Latino community using the conceptual framework of transitory identity, as developed by sociologist Stuart Hall (1994) in his chapter "Cultural Identity and Diaspora." This analysis alternates between important childhood and adolescent memories that shaped the formation of my black identity, and I rely principally on the medium of film to make these comparisons. The ultimate goal of this chapter is to illustrate the connections between Afro-Brazilians and African Americans and to broaden the discussion about the issues that should be addressed as global problems that are common to the entire African diaspora, propose solutions that can be replicated, as well as facilitate an ongoing dialogue between African descendants.

Films, Dreams, Identity, and Imaginary

The year was 1999, and I was 15 years old. I remember clearly the moments when I would be standing in the kitchen with my dad and upon hearing the opening soundtrack of the television series *The Fresh Prince of Bel Air*, we would come running to the living room to watch the episode. The lyrics began "In West Philadelphia born and raised." Up until that point, it was only through this music that I had any semblance of an idea about the city of Philadelphia.

One of the television channels in Brazil showed *The Fresh Prince of Bel-Air*, followed by the show *My Wife and Kids*. For my dad and me, it was like a family ritual to eat lunch and watch these two television shows every day. While watching *The Fresh Prince of Bel-Air*, I dreamed about the teenager Will Smith, black and unattached, and I wondered whether he might be interested in a girl like me, since in my own school none of the guys, who were largely white guys, seemed to be interested. I looked at his cousin in the show, the character of Hillary Banks, and imagined that one day, I would stop being (what I thought was) ugly and finally turn into someone as beautiful as she was. This was how the television was influencing my imagination.

I grew up in the 1990s, and was part of the Xuxa generation. Xuxa was a children's program host of a television show entitled "Queen of the Little Ones," and she became the model of beauty in Brazil (Simpson 2010). Xuxa had her sidekicks on stage known as the famous "Paquitas" and the dream of every child of my generation was to be one of them. There was just one small problem: Xuxa and all of the Paquitas were white, blonde, or, rather, they had European features. There was no way this would work for me since my hair was not straight, my nose was not thin, and my eyes were not blue. In fact, becoming a Paquita was nowhere near a possibility for me or for any other black child in the country. I never heard any formal explanation about the absence of black Paquitas; it was simply the norm in Brazil that the main characters had to be white.

The fact that I was in one of the only black and interracial families in the middle-class neighborhood where I lived in São Paulo had consequences for my school life. In the private schools that I attended, which were financially inaccessible to many Afro-Brazilians, I was one of the only black students in the classroom. The difference between my skin color and that of the other students was enough to elicit racist jokes in the classroom and it created in me a sense of being different that has continued to follow me through adulthood.

These experiences of my early school life explain the context in which my interest in *The Fresh Prince of Bel-Air* developed. The reality is that until the 2000s there was a significant absence of soap operas and television shows with black actors as the main characters in Brazil. While things have improved, they are still far from ideal. A watershed moment was the soap opera *Xica da Silva*, starring Taís Araújo. Broadcasted by the now defunct São Paulo station, TV Manchete, the novela aired in 1996 and was the first to star a black pro- tagonist. This means that 12 years had passed before I would ever see a black actress play the role of a protagonist. Unfortunately, in spite of having its high points, the novella widely exploited the sexual image of the character and included exaggerated violent scenes involving enslaved people. The song "Only God Can Judge Me" (Natasha Records BMG, 2002) by MV Bill revealed the insistence on the part of the media to reinforce the submission of blacks: "The Slave soap-opera is what the stations like, showing blacks whipped on their backs." In the book *A Negação do Brasil*, researcher and film critic Joel Zito Araújo (2000) explains in detail the origin of each black stereotype common on television. By rule they are submissive, hypersexualized, with disorganized families, and with the idea that interracial marriage is the salvation for racial progress.

Besides *The Fresh Prince of Bel-Air*, the film *Malcolm X* (Lee, 1992) by Director Spike Lee was another work that greatly influenced me. Seeing the life of this activist, his overcoming, and the legacy that he left was akin to a cathartic experience. It made me realize that the discourses that had been presented to me through the media were partial (and ideological); ultimately, blacks were not merely enslaved persons, subalterns, and ignorant of the history of their humanity as I commonly read in books and watched on television in Brazil.

Black communities were and are possessors of much knowledge; on the one hand, they were enslaved in a devastating way, but they always resisted the colonizers. As a result of this history, their socioeconomic status and continuing racism when compared with other groups have not managed to attain the same social development. I realized that there was much more to be learned about African ethnicity and African descendants, and it became my mission to research and become a voice of that history. This is what led me to my interest in films; it was a way that I would finally be able to see people who were similar to me, where people with my skin color and hair that looked like mine could live and act like everyone else, and contribute something positive.

Outside of Africa, Brazil is the country with the largest number of blacks in the world, totaling over 97 million. According to the 2010 Brazilian Census, the Brazilian population is 44.3 percent pardo (mixed), 7.6 percent preto (black), and 48 percent white. There are also other ethnic groups including indigenous and Asian people. Brazil is, therefore, a diverse country but this diversity is not reflected in media representation (Araújo 2000). On the Xuxa show, there was never a place for a black Paquita and this absence carried a huge weight because in Brazil, TV Globo was the television channel that everyone watched. Besides this, on no other shows were there any black hosts. This is why American shows and films that had black characters were so important to me. They filled the gap left by the media. *The Fresh Prince of Bel-Air* was a rare opportunity to see a stable and happy black family; these were my moments of pleasure and even strengthened our family bond.

Even though it was a foreign production, I felt that we could recognize ourselves as blacks and identify with those characters.

One of the direct consequences of the lack of representation of blacks in the media, in books, and in leadership positions in businesses is low self-esteem of the black population, mainly of black women who find themselves forced to meet unattainable standards of beauty that require them to straighten their hair, lighten their skin, and attempt to thin out their nose. I felt the same way as the other girls did: inferior. I focused on earning good grades and becoming a good athlete to socially compensate for my "ethnic disadvantage" (Hordge-Freeman 2015b). Given the Brazilian reality and the fact that the majority of blacks experience a process of identity formation, it is ultimately an identity that is constructed. As researcher and psychologist Neusa Santos Souza affirms, "to be black is not a given condition, a priori, it is coming to be. To be black is to become black" (1983, 77). To learn to appreciate one's identity is a slow process, and I have discovered it is a common experience throughout the African diaspora. Watching African American films and shows did not simply have an impact on how I saw myself as a black woman, but it also awoke in me a curiosity about the black community in the United States.

Obviously, the media portrays the United States as the land of opportunity, of consumption, of nice cars, and million-dollar homes. However, my references also included a number of films directed by blacks that told stories about African Americans including *Love & Basketball* (Gina Prince-Bythewood, 2000), *The Great Debaters* (Denzel Washington, 2007), *Jumping the Broom* (Salim Akil, 2011), *American Gangster* (Ridley Scott 2007), and *Men of Honor* (George Tillman Jr., 2000). Watching these films left me with a sense that the United States was the country in which blacks were free.

In addition to films, I have also been influenced by black music, as music genres such as R&B and hip-hop were commercially distributed in all of the nightclubs in São Paulo by the end of the 1990s, during my adolescence. At home, my dad also introduced me to vinyl records and the soul music of Marvin Gaye, Diana Ross, the Jackson 5, Stevie Wonder, and Ray Charles. In middle school, in the process of searching for my identity, African American figures such as Martin Luther King and Malcolm X represented for me the ultimate expression of what an organized black struggle could accomplish. From these examples, I had concluded that the United States was synonymous with a nation in which blacks had achieved more rights than in Brazil. My idea was that problems of belonging/acceptance, the dismal romantic life of black women, and the difficulty accessing the job market occurred much less in the United States than it did in Latin America.

Considering that the United States is the country of the Black Panther Party and the civil rights movement, I imagined that I would find a widely politicized black community there. In this utopia, black Americans would be seen as direct descendants of these movements, conscious of racial disparities, and ready to go into the streets when the black movement called. But what surprise did reality hold for me?

Arriving at the City of "Brotherly Love," Diversity in the City and Campus

Ever since I graduated from high school, I dreamed of studying film in the United States. I completed my bachelor's degree in radio and television at the Faculdade Cásper Líbero and looked for a graduate program in the country that appealed to me. After years of research, financial aid, and proficiency exams, I was accepted in the Master of Fine Arts (MFA) program at Temple University. It was a three-year program with the goal of either preparing students to continue an academic career or acting in the film industry. I chose Temple because of its strength in producing independent films that were socially relevant.

Another determining factor of my choice was its location. Temple was in Philadelphia, endearingly called Philly, meaning the city of fraternal love. According to the US Census, Philadelphia has over 1,525,006 inhabitants, making it the largest city in Pennsylvania and the second largest city on the east coast and the fifth most populous city in the United States. In 2013, Census data suggested that 44.2 percent of its population comprised blacks, 45.5 percent whites, and 13.3 percent Hispanics or Latinos.[1] Philadelphia is a demographically diverse city where African Americans, whites, black Africans, Caribbeans, and Latin Americans can be seen sharing the same social spaces. Studying in a city with a large black presence was an important decision because my objective was to live among African Americans.

Philadelphia has a more diverse population than São Paulo, which is the fifth largest city in the world. It is also the largest financial center in South America, and its metropolitan area includes over 20 million people according to the 2010 IBGE. In terms of race, São Paulo is 60.64 percent white, 30.51 percent brown (mixed-race/pardos), 6.54 percent black, 2.19 percent Asian, and 0.12 percent indigenous. Numerically and geographically, São Paulo is much larger than Philadelphia, but the black population is still around 40 percent in both of these cities. In the Brazilian case I use the term black to include mixed-race individuals and blacks. Encountering so many different cultures in Philadelphia was not such a shock to me because, in addition to identifying as Afro-Brazilian, I believe that the term Afro-Latina better reflects how I identify myself. I was born and raised in Brazil, but my parents are Peruvian, and one of them is Afro-Peruvian. Therefore, I consider myself

Afro-Peruvian-Brazilian. Having had access to Peruvian culture made me feel much more connected to Latin America than my Brazilian friends. The advantage of identifying as an Afro-Latina is that I identified with Brazilians, Caribbeans, and Latin Americans.

In Philadelphia, the black population is comprised of African Americans and also Africans who make up about 50,000 inhabitants representing different countries, including Ethiopia, Somalia, Sudan, Liberia, and Sierra Leon.[2] The majority of them arrived after 1980, as refugees, and they are second generation. Further adding to this ethnic diversity is the religious diversity. Among the three cities that I visited—New York, Washington, and Philadelphia—all have considerable black populations; however, Philadelphia has a greater population of black Muslims. It is easy to find black women covering their heads with veils or using a burqa. Many African Americans converted during the 1960s after being moved by Malcolm X and the doctrine preached by Elijah Mohammed in the Nation of Islam. In the last years, the number of African Americans who are Muslims in Philadelphia continues to grow, making this the metropolitan area with the highest percentage of Muslims in the country. There are more than 200,000 Islamic mosques in the city and 85 percent are predominantly African American.[3]

It is a city that is rich with many parks, museums, and cultural activities, but a large part of these are concentrated only in certain areas of the city. The city is also characterized by urban segregation, which is the consequence not simply of the times when whites and blacks circulated in different spaces, but also of racist housing practices and discrimination (Massey and Denton 1993). Philadelphia is the fifth most segregated state in the United States.[4] Blacks from a lower socioeconomic class largely live in North Philly, alongside Latinos from Puerto Rico. Blacks from the middle class are concentrated in West Philly, alongside the population of immigrants from Liberia and Ethiopia, among other places. However, in South Philly and the suburbs, the areas located on the periphery of the city, the large part of the population is white. Generally, the segregation is evident. During the two years that I lived in Philadelphia I had two distinct experiences. In the first year, I lived close to Temple's campus, in North Philly, in the area that is considered part of the "ghetto," and in the second, I lived in South Philly, which is considered a nicer area where upper- and middle-class families live.

My first impression of living in North Philly was like being on the set of *Do the Right Thing* by Director Spike Lee, where a population of marginalized black and Latin American families lived. Some homes appeared to be abandoned, dimly lit areas, with black unoccupied youth spending a good part of the day in the streets or involved in drugs. Another factor in these areas was poverty and the presence of Latino evangelical churches and African

American Methodist churches. I remember the taxis would refuse to drive me home and sometimes I had to wait for an hour and only after arguing with taxi drivers would they ultimately drive me home. University colleagues suggested that I move out of the neighborhood, because they hinted that I would be in danger if I continued to live there. But my goal was to really understand the black American community and, for this reason, I was determined to remain in the neighborhood. In retrospect, my desire to live in this community reflects a one-dimensional image of what black American life is, as I later learned that there is not just one authentic black experience.

In this neighborhood, there was a lack of recreational options, there were no parks, just basketball courts, and there were no theaters or cafés. Public maintenance and neighborhood planning were of low quality, including on snowy days, which greatly limited people's movement from one place to another. In Philadelphia, I observed firsthand that poor regions exist in the United States, as 12 percent of Philadelphia's population lives in extreme poverty, which is double the national average.[5] One observation that I made about poor neighborhoods in North Philly is that despite being poor, people generally still had access to commodities such as cars and household appliances. This is not the case in underdeveloped and still developing countries such as Brazil.

As is the case with all marginalized neighborhoods, there was an active social scene and music was central: I always heard passing cars blaring salsa, bachata, and musical styles common to Latin America. This same music could often be heard playing through the windows in the neighboring homes. In the summer, there were block parties, where the community closed streets and put out plastic pools for the children to have a party. These parties reminded me of the funk dances that took place in the periphery of São Paulo. There we also closed the streets in preparation for a party that would last all night. I also remember seeing the same plastic pools called "laje" in homes in the periphery of São Paulo and Peru. In the summer in Brazil, in every neighborhood one could hear music playing loudly, and I remember my first summer in my new neighborhood when I listened to the classic soul songs of the 1980s. I vividly remember my dad's old records at home—the music, the instruments, and the beat of the drum. It is remarkable the way that African culture reinvented itself in every country in the African diaspora. In Brazil, we have the *pandeiro* (tambourine), and in Perú (my parent's homeland) its equivalent is the *agogô*. We have the cajón, the cajita, and the zapateo afroperuano, and in the United States, there is also the zapateo, known as tap dancing and it is from this tradition that soul and rap were born.

One day I heard the sound of drums and whistles, and this reminded me of the sound of samba schools in Brazil. Was there a samba school in North

Philly? I tried to find out where the sound was coming from. I discovered it was the local drill team, called the "North Philly Dazzling Diamonds Drill Team," which offered drills as a cultural activity to neighborhood children, and sometimes practiced in the streets. This sense of recognizing something familiar in something that is foreign only reinforced the idea of our membership in the African diaspora, sharing the same roots. In each country where enslaved Africans were transported, we recreated our culture in distinctive, but also similar, ways.

In South Philly the experience was just the opposite. In part of the downtown area, the most touristy area, the infrastructure is of higher quality, there are more supermarkets and entertainment options, urban planning is more efficient, and there are more transportation options. However, there are notably few black residents in this region of the neighborhood because of racism (Countryman 2007). I experienced this first hand when I used my credit cards in stores in the area, and when I was asked if I lived in the neighborhood. This type of veiled or subtle racism was familiar and very typical in Brazil—I have known it since my childhood.

Also, in Philadelphia, in the richer areas, there were no protective gates or fences. Even in the suburban homes, there were often no bars between the entrance to the house and the sidewalk. In São Paulo, it was inconceivable to see an upper-class neighborhood with no gates or sophisticated security systems, cameras, and guards who protect the peace of the residents. Despite the doubtless inequality in the United States, the level of inequality was relatively less than in São Paulo. Of 128 countries that were evaluated on a development index completed by the Human Development Department of the United Nations Organization, the United States ranked fifth and Brazil seventy-ninth. In a report disseminated in July 2010, Brazil had the third worst index of inequality in the world.[6]

In contrast to my experience in São Paulo, in Philadelphia, I saw blacks circulating in the same areas that were frequented by people of the middle and upper class. In Brazil, despite not having a segregation law, we always had spaces marked by the absence of black people, and in many upper- and middle-class neighborhoods there is still restricted access to the black population. In the periphery, a large proportion of the population is of African descent. The shock that I had was less on finding that poverty existed in the United States. The more significant issue was confirming that here, in the country with a black president, where there was a considerable black middle-class community, and the land of the civil rights movements, unfortunately, the poorest neighborhoods were occupied by blacks and Latinos. All of these issues are linked to both historical and contemporary inequalities that created the same problems in all the regions of the diaspora: lack of basic infrastructure, housing shortages, poor schooling, unstructured families, and drugs.

With regard to the latter, it was disheartening to speak with young people from North Philly who did not have a sense of the future and who seemed to be disengaged from politics or even local issues. I had imagined that in every black community, I would find the sons and daughters of the Black Panther Party, people who were vigorously in search of their rights, but the reality was not like that at all. Yet, in the light of recent protests regarding racism and police brutality in the United States related to the police killings of Sean Bell, Trayvon Martin, and a slew of other unarmed black men, these organized protests led largely by black community organizations suggest that the legacy of the civil rights movements still remains a vibrant element of black protest in the United States.

I had very positive experiences in relation to black academics at Temple University. Franklin Cason, who lectured on the theory and politics of cinema, was my mentor in my second year in the program. His support and knowledge about black documentarians was fundamental for the development of my work. Cason and I had the type of mentor–student relationship that I had never experienced in São Paulo, as generally professors who advised my work were white and their expertise was not in the same area as my subject. Black students must often rely on outside mentorship in order to receive the guidance they need (Noy and Ray 2012). Outside of my department, black professors in other departments such as Sociology at Temple were also receptive and interested in developing partnerships with me.

Another factor that brought me to Temple was the fact that it has a strong nationally and internationally renowned African American Studies Department, directed by one of the pioneers of Afrocentric thought, Molefi Asante. It was through semester events that were organized by this department that I had exchanges with other black students on campus. Asante is also the founder of the MKA Institute and was happy when we met at an event. He excitedly introduced me as a "sister" who was both from Brazil and from Temple. He spoke about his mentor, Abdias do Nascimento, one of the most important black Brazilians in the struggle for racial equality in the country. Nascimento was an active member of the Pan Africanist International Movement and for several years he served as a visiting professor at US universities including Yale School of Drama (1969–1971), University of Buffalo, and The State University of New York. I was impressed with how many US academics were already aware of his trajectory since, in Brazil, his contributions were underappreciated (Nascimento 1989).

Generally speaking, I have found more opportunity and support from academic institutions in the United States than in Brazil. There is more space for black-centered films and personal narratives, rather than simply replicating traditional Hollywood styles. In Brazil, in all of my years of study, from primary to university, there were few black professors and the same goes for the number of black students. Certainly having professors and colleagues of the same racial background creates a more inviting environment and leads to the possibility of finding mentors who share your point of view. Research suggests that these points of similarity may often, though not always, lead to more supportive and effective relationships (Wolf-Wendel and Ruel 1999). Though ethnic or racial similarity alone is obviously not sufficient to establish dialogues, it can provide a base to serve as the beginning of a fruitful relationship.

Mobilization and Police Violence against Blacks in the United States and Brazil

Police violence against blacks is a problem in both the United States and Brazil. I witnessed media stories of police violence several times during my short stay in the United States. As an activist in Brazil I am acutely aware of police violence and have attended marches protesting such violence in São Paulo. Obviously racism has not been completely eradicated in American society and this is proven with persistent inequality in housing, healthcare (LaVeist 2005), black business ownership (Kopkin n.d.), and in cases of police brutality such as that of Trayvon Martin who George Zimmerman thought appeared "suspicious." This is common parlance used to describe black Americans. Interestingly Afro-Brazilians are also viewed as suspicious and are often targets of police violence (Alves 2013; Smith 2014). There was also the case of the 17-year-old boy, Jordan Davis, who was killed because he was listening to rap music in his car and a white man thought it was too loud. The most recent case was of Freddie Gray, a young 25-year-old African American man, who was a victim of police violence on April 19, as a result of which his spine was fractured and he later succumbed. Racism exists in the United States but alongside these high-profile cases there are serious discussions of racism. In all these cases African Americans mobilized, and in the case of Trayvon Martin, African Americans mobilized throughout the country. Although it seems as though in everyday life black youth in Philadelphia are not very politically active, these cases made national news and led to African Americans mobilizing throughout the country.

Similarly in Brazil, the problem of police violence against young blacks from low-income backgrounds has reached stratospheric numbers and we have also mobilized thousands of people in the streets such as in the Second International March against the Genocide of Black People, which took place on August 22, 2014, in the city of São Paulo. According to the military police, at least one thousand people participated in the march. Other protests were scheduled to occur in 18 other Brazilian states and in 15 countries. Jaime Alves (2013) discusses the efforts of black women who organized an event to honor the children who have been slain by the police throughout São Paulo. This type of organization is powerful in reclaiming the dignity of black youth.

Although police brutality is a problem in many countries in the African diaspora, it is interesting how, in my experience, many black Americans ignore that this is a global problem. I still continue to hear many people who believe in the myth of racial democracy because of racial miscegenation in Brazil. This ideology was propagated by Gilberto Freyre in his famous book *The Masters and the Slaves* (1986) and it continues to influence perceptions of Brazil as a racial paradise. In his book *Mixture or Massacre* (1989) Abdias do Nascimento challenges this false idea. Yet some people still think race relations in Brazil are not an issue when compared to racial problems in the United States. As an African-descended activist I believe it is important that African Americans and Afro-Brazilians know that the issue of police brutality is a global problem and that we must find ways to fight against such pernicious acts.

"Making Diaspora" with Afro-Latinos and African Americans

I remember my initial seminar at Temple University where a professor talked about a group of alumni and the racial terms used to describe them were black, white, Asian, multiracial, and Hispanic. The black category was used to describe African Americans. In the United States, black Latinos are considered an ethnic group within the term Hispanic and this surprised me. This designation for Afro-Latinos makes it difficult to form a black identity for Latino immigrants and limits unity between all African descendants who live in the country. In conversations with students from the Dominican Republic whom I would consider racially black, they told me they did not consider themselves black because this term implies that they would be African American, even though one student admitted she knew she had African ancestry. However, in the United States she is considered Hispanic.

This has generated significant discussion about the origin and potential significance of the term Afro-Latino. If, according to the US national census, blacks comprise only 13 percent of the population, what would the number be considering Afro-Latinos? Another point is that this schism generates resistance from nonblack Latinos to accept them as Latinos. However, in acknowledging African ancestry the term Afro-Latino allows African Americans to understand that there are blacks in other parts of Latin America. To respond to these issues, Miriam Jiménez Román and Juan Flores wrote *The Afro-Latin@ Reader: History and Culture in the United States* (2010), which is an edited volume that includes contributions from Afro-Latino scholars and activists.

In Peru, a young black activist I interviewed told me "We have to make the diaspora." I believe that my job to teach people about the African presence in Brazil is very important. Africans and blacks throughout the diaspora need to unite to understand that we all face similar realities. Scholars such as Tanya Golash-Boza (2012) find that in Peru there is minimal memory of slavery and, when slavery is mentioned, reveals the diverse ways that African descendants understand and interpret their identity and their connection to a history of slavery. Nevertheless, it is important that we all know about these complicated histories and present-day realities. "Making diaspora" as an activist working on films that focus on African descendants will enable us to connect with each other to build stronger communities with the goal of full liberation for all those in the African diaspora.

Notes

1. US Census: http://quickfacts.census.gov/qfd/states/42/42101.html. Accessed on June 2, 2015.
2. Available at http://globalphiladelphia.org/communities/african. Accessed on December 2, 2014.
3. Available at http://weeklypress.com/islam-in-the-city-of-brotherly-love-p531-97.htm. Accessed on December 2, 2014.
4. http://atlantablackstar.com/2014/03/24/10-of-the-most-segregated-cities-in-the-u-s/6/
5. http://articles.philly.com/2014-09-26/news/54322611_1_deep-poverty-poverty-line-south-philadelphia
6. See http://hdr.undp.org/en for additional statistics about Brazil's performance on the Human Development Index.

Conclusion: Toward a Future African Diasporic Approach to Research Diaspora

Gladys L. Mitchell-Walthour and Elizabeth Hordge-Freeman

B lack transnational engagement between researchers in Brazil and those in the United States adds one more layer to the "major dialogue shaping the cultures and politics of the Afro-Atlantic world" (Matory 2006, 153). Contrary to the notion that intellectual trends are guided by the whims of the "invisible hand" of the academy, there are cognitive orientations and perceived cultural commonalities that explain the origin and persistence of black researchers' interests in their counterparts in the United States and Brazil. Beyond serving as a logical point of comparison, due to similar (though not identical) histories, their sense of shared political goals, racial commonalities, and solidarity against racism means that diasporic citizens engage in dialogues to monitor, analyze, and refine movements and programs in their own countries (Pereira 2013).

Broader Research Implications of Positionality

Most of the contributions to this volume are by researchers and/or activists who view their positionality as a type of resource. Rather than simply superimposing their ideas about race relations onto the countries they visit, the narratives and experiences that these researchers recount illustrate that they are willing to transform their racial schemas and conceptualizations of race in order to recognize the significant differences that exist. Some of the researchers drastically alter perspectives about themselves stemming from the unique experiences that their multiple insider/outsider identities afford them (Chapter 5 by Joseph illustrates this well). In this volume, the contributors' highly personal accounts of eye-opening experiences, which they respond to with embarrassment, confusion, and/or anger, reflect challenges and opportunities for the transnational researcher. Their studied sensitivity to the uncomfortable moments, which they use to reflect and reevaluate their expectations and impressions, leads to a nuanced examination of race relations in both Brazil and the United States. These moments are made clear by the misrecognitions discussed in most chapters, but are portrayed particularly saliently by Reighan Gillam, Chinyere Osuji, and Gladys L. Mitchell-Walthour.

In addition, African American researchers conducting research in Brazil now see themselves in very different ways. Any good researcher should question how their positionality shapes their experiences in the field and how it inhibits or limits the information they get in the field. How does one's positionality impact who gets interviewed? How does it shape one's access to certain places and certain discourses? In what ways do experiences in the field improve or perhaps complicate the relationships in the communities that are studied? These questions are relevant to all researchers, but for black researchers in Brazil or in the United States who often operate in environments where blackness is viewed as inferior, fieldwork can be anxiety-inducing. We hope that the chapters here allay some of this angst by offering potential strategies for dealing with uncomfortable and offensive situations. Perhaps the best example of this is Lúcio Oliveira's experience where he uses strategic silence when he is profiled by a police officer on UCLA's campus. Another example in this book is Jaira J. Harrington's use of her constant misidentifications to explore the "aesthetics of power."

Race and the Politics of Knowledge Production is relevant at this precise juncture because the interactions between black researchers in the United States, Brazil, and other diasporic countries are expanding rapidly. Affirmative action policies in Brazil have played an important role in increasing the number of Afro-Brazilians who graduate from college and this should create a pipeline of students pursuing doctoral degrees. We know that their presence alone will not be sufficient, especially given Mojana Vargas's analysis of the practices and structural inequality of Brazilian institutions. Nevertheless, visibility and access are the first step, and Afro-Brazilian men and women graduate students are studying a number of topics and a number of them are studying subjects relevant to Afro-Brazilian communities. Flávio Thales Ribeiro Francisco, a doctoral student at the University of São Paulo, has studied the construction of race in the *Clarim da Alvorada* (1924–1932), a black newspaper in Brazil. Flávia Rios is a doctoral student at the University of São Paulo studying race relations, social movements, and public policy. She coauthored the book *Lélia Gonzalez* (2010), which is a biography of Gonzalez who was a black Brazilian activist and scholar. Jaqueline Lima dos Santos, a doctoral student at the University of Campinas, is conducting research on ethnoracial relations, gender, and education in Brazil and Portuguese-speaking African countries. Within this edited volume, doctoral students Lúcio Oliveira and Daniela Gomes also research exciting topics pertinent to the Afro-Brazilian community and the African diaspora at large.

For many of the Brazilian contributors who appear in Part III, their introduction to African American culture through soul or R&B music and television shows was instrumental in constructing an idealized image about the daily life and revolutionary spirit of African Americans in the United States. In Chapters 10 and 12, Gabriela Watson and Daniela Gomes, respectively, reflect on some of the disappointments they faced in Philadelphia and Atlanta, as they discovered that this image was a mirage. Lúcio Oliveira enters the United States with different expectations, partly based on his previous experience in the Race and Democracy Project. While his conceptions of the United States are different (and critical), his personal experiences are strikingly similar to the commentaries by Gomes da Silva, Aurazo, and Vargas, who relate the way that aesthetics, self-esteem, and racial invisibility created an emptiness that is addressed through the racial activism, aesthetics, and possibilities of the "African American way of life." Even though we recognize the power that these scholars gained from these US images, there are important questions about the particular way that blackness is constructed and packaged for international audiences. There is something very inspiring, and disturbing, about how US images so thoroughly shaped the contributors' lives, leading them to seek out not just the country but even particular cities that aligned with their image of racial radicalism. All the contributors generally agree that the outcomes were positive, but we must problematize the way that "hegemonies of diaspora" might tend to reflect and reinforce broader cultural and national hierarchies (Thomas 2007).

In contrast to the way Brazilians encountered black American culture, African American researchers encountered Brazil through the academy or academic-related events. The economic and cultural position of the United States in the world leads to unequal power relationships that shape these researchers' access to education and information. African Americans are in a way restricted to potential dialogue with Afro-Brazilians unless they are in the academy. Therefore, while the United States exports black American culture readily (both jazz and hip-hop music are examples), Afro-Brazilian music is not as common there, not least because relatively few people in the United States know people of African descent. These power asymmetries between black transnational researchers become clear through the Brazilian contributors who note that in their interactions Americans are totally ignorant about

Brazil, have internalized racist and sexist ideas about Brazilian sexuality, or believe that Brazil is a racial democracy.

Institutional and Organizational Allies

In terms of creating sustained and meaningful dialogues, it is through the encounters of African American and Afro-Brazilian researchers that we observe some of the most sustained cultural and intellectual exchanges that impact research and knowledge production. In Chapter 2, David Covin illustrates the potential to forge strong transnational alliances in ways that can bridge black scholars from Brazil and the United States. Indeed, this collaborative project led to Afro-Brazilians' entrance and graduation from competitive doctoral programs, creating both a pipeline of scholars and a model of what a collaborative diasporic initiative might entail. While the Race and Democracy Project no longer exists, the outcomes of the project have produced sustainable results as members of the network still provide mentorship and continue to support the goal of professionalizing future researchers and activists. Moreover, conference panels such as those at the National Conference of Black Political Scientists, and many others both nationally and internationally, provide a space for past mentors to continue to meet and mentor one another.

Today there are no projects similar to the Race and Democracy Project in the United States with the purpose of bringing together Afro-Brazilian scholars and training Afro-Brazilian graduate students. This could be due to the difficulty in funding such projects in a climate where social science funding has been under attack. For example, in 2013, the National Science Foundation (NSF) temporarily cut political science funding for projects that were not aimed at national security interests or economic development. Considering that the Race and Democracy Project received a grant from the NSF justifies that the project led by political scientists would be particularly challenging. Nonetheless, the Abdias do Nascimento Academic Development program sponsored by the Brazilian Ministry of Education, introduced in 2014 between Brazilian and foreign universities, supports undergraduate and graduate studies for black, brown, and indigenous Brazilian students. This exchange program is a very recent example of transnational engagement that will help promote the type of involvement that will be beneficial to knowledge production and cultural exchange.

As universities cut funds for large multimillion-dollar programs, students actively pursue other ways to engage in transnational and global opportunities. Rather than viewing these programs as income generators, faculty and institutions are challenged to find ways to use these programs to provide mutually beneficial opportunities for students and diasporic communities. While launching a study abroad program is not an option for most faculty, introducing global awareness by integrating comparative perspectives into the course curriculum can effectively engage students in these questions. By raising global awareness and learning concrete examples of how course curriculum impacts and potentially improves the lives of others, students will come to appreciate the interconnectedness of societies, and invest more in supporting the integrity of societies outside of their own. Although academic institutions can be beneficial, we know that not all programs are equally valued and supported. In Chapter 1, Kia Lilly Caldwell demonstrates the way black women's voice continued to be marginalized in scholarship and even within spheres where one would expect their inclusion. The field of intersectionality is an area of study that was significantly developed by Kimberlé Crenshaw, who is the founder and executive director of the African American Policy Forum. These organizations comprise activists, policy makers, and academics who might be particularly receptive to promoting the visibility, rights, and equality of black Brazilian women.

Looking Forward

We are pleased to have edited a collection that drew contributors from both the United Sates and Brazil. A future edited volume might pay attention to how the theme of the book is framed and to whom it will be disseminated. In this case, it will largely be disseminated to US-based libraries and audiences, and although this is a customary practice in the United States, it limits the accessibility of the book to Portuguese speakers (reinforcing some of the dynamics that we have critiqued here). We value our Brazilian readers and ideally would like this to be accessible to those working on issues related to race, class, and gender. So, a future volume should address this.

In terms of the overall participation in the volume, there is a significant representation of black Brazilianists who are women, but African American and Afro-Brazilian men are underrepresented. Additionally, we are cognizant that despite our efforts to include Afro-Brazilian scholars, most of those participating in this volume are doctoral students. These positions within the academy reflect power relations in the larger society. However, the fact that more Afro-Brazilians are in graduate schools now than in the past is promising, and should reflect in future edited volumes that address these questions of positionality.

In a similar vein, a future edited collection should ideally include coeditors representing the United States and Brazil. Although two African American women organized this volume, we hope that in the future Afro-Brazilians will take the lead in organizing a volume. This would signal the type of transnational dialogue we advocate in the book and reflect the growing visibility and challenges facing Afro-Brazilian scholars in the field. Because of black women's particular experiences, we may be more likely to feel committed to such a project despite the many academic and personal pursuits we have. Black women feminists, both scholars and activists alike, have often felt the burden of being committed to the community and scholarship. The Combahee River Collective, which was made up of black women scholars and activists, is but one example of an effort to combine both activism and knowledge production. Historically the burden of public and private obligations for black women is known. Sojourner Truth's well-known "Ain't I a Woman" speech delivered in 1851 at the Women's Convention in Ohio is one of many examples of black women as physical laborers as well as mothers (Painter 1997). She notes that she did as much work as men and also bore children. Despite the fact that rights were denied to her as a black woman, she reclaimed her dignity. Black women scholars, like other women scholars of color, face unique challenges in the academy and they are pulled in many directions as keepers of the community and beholden to work outside the personal sphere.

Gutiérrez y Muhs et al.'s ground-breaking edited volume *Presumed Incompetent* (2012) is an excellent work that details both the challenges women scholars of color face and tools of survival. Black women academics carry the double burden of ensuring that the voices of black people throughout the diaspora are heard through their intellectual activism while remaining committed to personal and academic pursuits. It is not simply an intellectual exercise but rather it is considered by many to be our responsibility to provide support to researchers who are also working to promote equality in local and global communities.

Sustained Dialogue

Essential to knowledge production is that it is not produced in a vacuum but there should be sustained dialogue with other academics. Both Caldwell's and Covin's works serve as examples of how African-descended scholars can collaborate on projects that actually lead to empowerment of African descendants. There are challenges that African-descended scholars might face when trying to create sustained dialogue. Social media can help fill this void in some ways through blogs, Facebook, and other forums that create opportunities for engagement in each other's work as well as reaching out to one another. In fact, it was because of Daniela Gomes's well-written and important blog about the murder of an innocent Afro-Brazilian woman, Cláudia Ferreira, by Brazilian police posted on a friend's Facebook page that we learned that she was studying in the United States. It was through social media that we became aware of her work and used this forum to invite her as a contributor. These forums can create opportunities for sustained dialogue between Afro-Brazilians and African Americans.

In the future, black researchers in both countries should create opportunities for interaction within their respective countries of research. Some of these opportunities naturally present themselves during conferences that may include sessions on topics relevant to both groups. Some of the participants in this volume have made presentations at the National Conference of Black Political Scientists. Another example is the Brazil Studies Association conference and the Latin American Studies Association. Both host conferences where scholars have organized panels addressing racial inequality and exclusion of Afro-descendants. Conferences specifically focusing on Afro-descendants have become more common, which also creates such opportunities. Yet a formalized effort to bring together African American and Afro-Brazilian scholars would be an important step in ensuring sustained dialogue.

Activism

Producing knowledge is important but is not the only goal of this volume. We hope that it serves as a project to engage scholars in active research that leads to empowerment on both sides. In Chapters 2 and 3, Covin and Hordge-Freeman, respectively, discuss empowering Afro-Brazilians, future Afro-Brazilian researchers should think about ways to empower African American youth. While the perception is that African American youth are already empowered and politically active because of the legacy of black activism, many African Americans are not aware of the challenges facing African descendants in other parts of the world nor are they aware of the type of organizing blacks in the diaspora have engaged in. Afro-Brazilian scholars find themselves in a unique position as they are from a country that historically embraced the notion that racism does not exist. Afro-Brazilian researchers in this volume have noted experiences of racial discrimination, thus they hold the unique experiential position of addressing commonly held beliefs by Americans that Brazil is free of racism. We must be creative about the types of liberating projects we can be involved in as scholars. These projects could involve African American youth, informing them about the current situation of blacks in Brazil, or they could be projects aimed at African American adults, educating them about Brazil beyond the tourist trap of visiting cultural sites. We all have a responsibility to fulfill as teachers through activism.

African American researchers have the responsibility of creating opportunities for Afro-Brazilians to participate in the Brazilian economy, which has greatly developed over the past 10 years. In Chapters 2 and 3, Covin and Hordge-Freeman, respectively, discuss how researchers can use institutions and programs to organize, engage, and train future scholars. Future programs should also focus on empowering Afro-Brazilians rather than simply involving them in programs as mere participants. An asset-based approach is the most viable approach at building the capacity of people to participate in a more global market. As black researchers we must critically think about small and big ways of contributing to the lives of Afro-Brazilian researchers and more importantly to Afro-Brazilians generally. Hiring Afro-Brazilian research assistants when conducting research is an important first step toward training future scholars. Contacting nonprofit organizations and others that may not be formally nonprofit but that are embedded in neighborhoods with the aim of improving the lives of Afro-Brazilians is an important practical step that we should take to ensure that we are not simply extracting information from communities but are critically engaged in those communities. Making these special efforts is at the heart of what it means to be a diasporic community.

Challenging Racism in Brazil and the Push for Post-Racialism in the United States

This book and our call for new ways of engaging diaspora come at an interesting historical moment. In the United States, while progressive-minded scholars continue to challenge racism, sexism, and classism, those in mainstream media are pushing the idea that America is post-racial. They cite the election of African American president Barack Obama as an example of a post-racial society where everyone has equal opportunity. As a result, many universities are being challenged for having an Ethnic Studies department. Universities that have affirmative action policies are being challenged and in some cases universities such as the University of Michigan, Ann Arbor, can no longer use race as a factor when considering admission. On the contrary, in Brazil race is becoming a more acceptable topic in the media and among citizens. In addition, the number of Afro-Brazilians enrolling in college is increasing due to the implementation of quota programs in university admission. This book gives us an insight into how both scholars and activists can challenge racism, sexism, and classism while empowering African descendants in these changing racial climates. America's racial climate is becoming similar to Brazil's racial climate of old where racial mixture was seen as the solution and the society was proclaimed to be nonracist. In 2013, *National Geographic* published "The Changing Face of America," where there is a discussion about the ambiguity of race and photographs that accompany the article. Lise Funderburg writes that "[w]e've become a country where race is no longer so black or white." Yet the Brazil case and the current status of police violence and killings against blacks in the United States, and protests as a reaction to these killings, show us that simply claiming that a nation is not racist and having racially mixed people does not mean that African descendants are not marginalized in society. Similarly, because many of the quota programs in Brazil have been discussed as temporary solutions in that they will be reevaluated in a few years does not mean that progress can be slowed as we have seen in the case of the United States. The experiences of Afro-Brazilian researchers in the United States and African American researchers in Brazil allow a real-time experience of how these dynamics work in both societies as both are evolving in terms of race relations. Researchers in both countries gain insights to challenge these dynamics and empower those in the African diaspora.

Bibliography

ACIDI—Alto Comissariado para a Imigração e o Diálogo Intercultural. "Inter-culturalidade. Uma realidade em Portugal?" *ACIDI REVISTA* 83 (August 2010). Lisboa. http://www.acidi.gov.pt/_cf/26882

Agard-Jones, Vanessa and Manning Marable. 2008. *Transnational Blackness: Navigating the Global Color Line*. New York: Palgrave Macmillan.

Alberto, Paulina L. 2009. "When Rio Was Black: Soul Music, National Culture, and the Politics of Racial Comparison in 1970s Brazil." *Hispanic American Historical Review* 89(1): 3–39.

Alberto, Paulina L. 2011. *Terms of Inclusion: Black Intellectuals in Twentieth-Century Brazil*. Chapel Hill, NC: University of North Carolina Press.

Alvarez, Sônia, Claudia de Lima Costa, Verónica Feliu, Rebecca Hester, Norma Klahn, and Millie Thayer, eds. 2014. *Translocalidades/Translocalities: Feminist Politics of Translation in the Latin/a Américas*. Durham, NC: Duke University Press.

Alves, Jaime. 2013. "From Necropolis to Blackpolis: Necropolitical Governance and Black Spatial Praxis in São Paulo, Brazil." *Antipode: A Radical Journal of Geography* 46: 323–339.

Amado, Jorge. 1962. *Gabriela, Clove and Cinnamon*. Translated by James L. Taylor and William L. Grossman. New York: Knopf.

"Americas Barometer." 2010. Ethnicity Module of the Project on Ethnicity and Race in Latin America (PERLA), Latin American Public Opinion Project of Vanderbilt University, www.AmericasBarometer.org

Anderson, Benedict. 2006. *Imagined Communities: Reflections on the Origin and Spread of Nationalism*. New York: Verso.

Andrioli, Antônio I. 2005. "O Lugar das Ciências Humanas na Universidade." *Revista Iberoamericana de Educación* 5(37): 1–15.

Ansell, Nicola. 2001. "Producing Knowledge about 'Third World Women': The Politics of Fieldwork in a Zimbabwean Secondary School." *Ethics, Place and Environment* 4: 101–116.

Araújo, Joel Z. 2000. *A Negação do Brasil: O Negro na Telenovela Brasileira*. São Paulo, Brazil: SENAC.

Araújo, Marta and Silvia R. Maeso. 2013. "The Absent Presence of Racial: Political and Pedagogical Discourses on History, 'Portugal' and (Post)Colonialism." *Educar em Revista* 47: 145–171.

Bailey, Stanley. 2009. *Legacies of Race: Identities, Attitudes, and Politics in Brazil*. Palo Alto, CA: Stanford University Press.

Bairros, Luiza. 1991. "Mulher negra: O reforço da subordinação." In *Desigualdade racial no Brasil contemporâneo*, edited by Peggy Lovell, 177–183. Belo Horizonte: MGSP Editores.

Bambara, Toni Cade and Elenora Traylor. 1970. *The Black Woman: An Anthology*. New York: Signet.

Barbosa, Lúcia Maria de Assunção, Silva, Petronilha Beatriz Gonçalves and Silvério, Valter Roberto, eds. 2003. *De preto a afro-descendente: Trajetos de pesquisa sobre o negro, cultura negra e relações étnico-raciais no Brasil*. São Carlos, Brazil: EdUFSCar.

Barros, Zelinda dos Santos. 2003. "Casais inter-raciais e suas representações acerca de raça." In *Filosofia e Ciências Humanas*. Salvador, Bahia: Universidade Federal da Bahia.

Beck, Hamilton. 1996. "W. E. B. Du Bois as a Study Abroad Student in Germany, 1892–1894." *Frontiers: The Interdisciplinary Journal of Study Abroad* 2(1): 45–63.

Bell, Karen and A. W. Anscombe. 2013. "International Field Experience in Social Work: Outcomes of a Short-Term Study Abroad Programme to India." *Social Work Education* 32(8): 1032–1047.

Birdwell, Sandra. 2009. "Negação e falta de representação: 'TV Negra' no Brasil e nos Estados Unidos." In *Retratos e Espelhos: Raça e Ethnicidade no Brasil e nos Estados Unidos*, edited by Vinicius Vieira and Jacquelyn Johnson, 245–267. São Paulo, Brazil: FEA/USP.

"Black Women of Brazil." Blackwomenofbrazil.com

Blalock, Hubert M. 1967. *Toward a Theory of Minority Group Relations*. New York: John Wiley & Sons.

Bonilla-Silva, Eduardo and David R. Dietrich. 2009. "The Latin Americanization of US Race Relations: A New Pigmentocracy." In *Shades of Difference: Why Skin Color Matters*, edited by Evelyn Nakano Glenn, 40–60. Palo Alto, CA: Stanford University Press.

Borges, Dain. 1993. *The Family in Bahia, Brazil, 1870–1945*. Stanford, CA: Stanford University Press.

Bourdieu, Pierre and Loïc Wacquant. 1992. *An Invitation to a Reflexive Sociology*. Chicago, IL: University of Chicago Press.

Bourdieu, Pierre and Loïc Wacquant. 1999. "On the Cunning of Imperialist Reasoning." *Theory, Culture, and Society* 16(1): 41–58.

Boyce Davies, Carole and Babacar M'Bow. 2007. "Towards African Diaspora Citizenship: Politicizing and Existing Global Geography." In *Black Geographies and the Politics of Place*, edited by Katherine McKittrick and Clyde Woods, 14–45. Cambridge, MA: South End Press.

Brown, Jacqueline Nassy. 2005. *Dropping Anchor, Setting Sail: Geographies of Race in Black Liverpool*. Princeton, NJ: Princeton University Press.

Buford May, Reuben. 2014. "When the Methodological Shoe is on the Other Foot: African American Interviewer and White Interviewees." *Qualitative Sociology* 37(1): 117–136.

Burdick, John. 1998. *Blessed Anastacia: Women, Race, and Popular Christianity in Brazil*. New York: Routledge.

Burdick, John. 2013. *The Color of Sound: Race, Religion, and Music in Brazil*. New York: New York University Press.

Bush, Rod, ed. 1984. *The New Black Vote: Politics and Power in Four American Cities*. San Francisco, CA: Synthesis Publishers.

Butler, Kim. 1998. *Freedoms Given, Freedoms Won: Afro-Brazilians in Post-Abolition São Paulo and Salvador*. New Brunswick, NJ: Rutgers University Press.

Caldwell, Kia Lilly. 2000. "Fronteiras da diferença: raça e a mulher no Brasil." *Revista Estudos Feministas* 8(2): 91–108.

Caldwell, Kia Lilly. 2004. "'Look at Her Hair': The Body Politics of Black Womanhood in Brazil." *Transforming Anthropology* 11: 18–29.

Caldwell, Kia Lilly. 2007. *Negras in Brazil: Re-envisioning Black Women, Citizenship, and the Politics of Identity*. New Brunswick, NJ: Rutgers University Press.

Campt, Tina and Deborah A. Thomas. 2008. "Gendering Diaspora: Transnational Feminism, Diaspora and Its Hegemonies." *Feminist Review* 90(1): 1–8.

Carby, Hazel. 1987. *Reconstructing Womanhood: The Emergence of the Afro-American Woman Novelist*. New York: Oxford University Press.

Cardoso, Abílio H. 1989. "A universidade Portuguesa e o Poder Autonómico." *Revista Crítica de Ciências Sociais* 27/28: 125–145.

Carneiro, Sueli. 1990. "A organização nacional das mulheres negras e as perspectivas políticas." *Vozes* 84(2): 211–219.

Carneiro, Sueli. 1995. "Gênero, raça e asenção social." *Estudos Feministas* 3(2): 544–552.

Carneiro, Sueli. 1999. "Black Women's Identity in Brazil." In *Race in Contemporary Brazil: From Indifference to Inequality*, edited by Rebecca Reichmann, 217–228. University Park, PA: Pennsylvania State University Press.

Carneiro, Sueli and Thereza Santos. 1985. *Mulher negra*. São Paulo, Brazil: Nobel/ Conselho da Condição Feminina.

Castro, Mary Garcia. 1993. "The Alchemy between Social Categories in the Production of Political Subjects: Class, Gender, Race and Generation in the Case of Domestic Workers' Union Leaders in Salvador-Bahia, Brazil." *The European Journal of Development Research* 5(2): 1–22.

Castro, Mary Garcia and Miriam Abramovay. 2006. *Relações Raciais na escola: reprodução de desigualdades em nome da igualdade*. Brasília: UNESCO.

Caton, Kellee and Carla Almeida Santos. 2009. "Images of the Other Selling Study Abroad in a Postcolonial World." *Journal of Travel Research* 48(2): 191–204.

Cavalleiro, Eliane dos Santos. 2000. *Do Silêncio do Lar ao Silêncio Escolar: Racismo, Preconceito e Discriminação na Educação Infantil*. São Paulo, SP: Humanitas, FFLCH-USP: Editora Contexto.

Charles, Ray. 2000. *Georgia On My Mind*. In *The Very Best of Ray Charles*, Vol. 1.

Childs, Erica Chito. 2005a. "Looking Behind the Stereotypes of the 'Angry Black Woman': An Exploration of Black Women's Responses to Interracial Relationships." *Gender & Society* 19: 544.

Childs, Erica Chito. 2005b. *Navigating Interracial Borders: Black-White Couples and Their Social Worlds*. New Brunswick, NJ: Rutgers University Press.

Chong, Kelly H. 2008. "Coping with Conflict, Confronting Resistance: Fieldwork Emotions and Identity Management in a South Korean Evangelical Community." *Qualitative Sociology* 31: 369–390.

Chow, Esther Ngan-Ling, Chadwick Fleck, Gang-Hua Fan, Joshua Joseph, and Deanna M. Lyter. 2003. "Exploring Critical Feminist Pedagogy: Infusing Dialogue, Participation, and Experience in Teaching and Learning." *Teaching Sociology* 31: 259–275.

Christian, Barbara. 1980. *Black Women Novelists: The Development of a Tradition, 1892–1976*. Westport, CT: Greenwood Press.

Christian, Barbara. 1985. *Black Feminist Criticism: Perspectives on Black Women Writers*. New York: Pergamon Press.

Clarke, Averil Y. 2011. *Inequalities of Love: College-Educated Black Women and the Barriers to Romance and Family*. Durham, NC: Duke University Press.

Collins, Patricia Hill. 1986. "Learning from the Outsider Within: The Sociological Significance of Black Feminist Thought." *Social Problems* 33(6): 14–32.

Collins, Patricia Hill. 1990. *Black Feminist Thought: Knowledge, Consciousness and the Politics of Empowerment*. Routledge.

Collins, Patricia Hill. 1998. *Fighting Words*. Minnesota: University of Minnesota Press.

Collins, Patricia Hill. 2000. *Black Feminist Thought: Knowledge, Consciousness, and the Politics of Empowerment*, 2nd ed., Rev. tenth anniversary ed. New York: Routledge.

Collins, Patricia Hill. 2004. *Black Sexual Politics*. New York: Routledge.

Contins, Marcia. 2005. *Lideranças Negras*. Rio de Janeiro: Aeroplano.

Cooley, Charles Horton. 1902. *Human Nature and the Social Order*. New York: Scribners.

Costa, Claudia de Lima and Sônia Alvarez. 2014. "Dislocating the Sign: Toward a Transnational Feminist Politics of Translation." *SIGNS* 39(3): 557–563.

Costa, Leeray M. and Karen J. Leong. 2012. "Introduction Critical Community Engagement: Feminist Pedagogy Meets Civic Engagement." *Feminist Teacher* 22(3): 171–180.

Countryman, M. J. 2007. *Up South: Civil Rights and Black Power in Philadelphia*. Philadelphia, PA: University of Pennsylvania Press.

Couto, Richard. 1991. *Ain't Gonna Let Nobody Turn Me Round: The Pursuit of Racial Justice in the South*. Philadelphia, PA: Temple University Press.

Couto, Richard. 1993. "Narratives, Free Spaces, and Political Leadership." *The Journal of Politics* 55(1): 57–79.

Covin, David. 1989. "On Race, Class, and Power in Brazil." *Western Journal of Black Studies* (Winter) 1: 152–153.

Covin, David. 1990a. "Afrocentricity in the MNU." *Journal of Black Studies* (Winter) 21: 126–144.

Covin, David. 1990b. "Ten Years of the MNU: 1978–1988." *Journal of Third World Studies* 7(2): 242–264.

Covin, David. 1997a. "Narrative, Free Spaces, and Communities of Memory in the Brazilian Black Consciousness Movement." *Western Journal of Political Science* 21(4): 272–279.

Covin, David. 1997b. "Social Movement Theory in the Examination of Mobilization in the Black Community." *National Political Science Review* 6: 94–109.

Covin, David. 2001. "Black Activists During the Ebb Tide of a Social Movement." *National Political Science Review* 8: 100–123.

Covin, David. 2006. *The Unified Black Movement in Brazil, 1978–2002.* Jefferson, NC and London: McFarland & Co.

Covin, David. 2009. *Black Politics After the Civil Rights Movement: Activity and Beliefs in Sacramento, 1970–2000.* Jefferson, NC and London: McFarland & Co.

Craig, Maxine L. 2002. "Race, Beauty and the Tangled Knot of a Guilty Pleasure." *Feminist Theory* 2(7): 159–177.

Cumberbatch, Prudence D. 2009. "Transnationalism and the Construction of Black Political Identities." *Radical History Review* 103: 163–174.

Damasceno, C. M. 2000. "'Em casa de enforcado não se fala em corda': Notas sobre a construção social da "boa" aparência no Brasil." In *Tirando a máscara. Ensaio sobre o racismo no Brasil,* edited by Antônio Sérgio Guimarães and Lynn Huntley. São Paulo, Brazil: Paz e Terra.

Daniel, G. Reginald. 2006. *Race and Multiraciality in Brazil and the United States: Converging Paths?* University Park, PA: The Pennsylvania State University Press.

Daniel, G. Reginald. 2010. *More than Black: Multiracial Identity & New Racial Order.* Philadelphia, PA: Temple University Press.

Davis, Angela Y. 1981. *Women, Race & Class.* New York: Vintage Books.

Davis, Angela Y. 1989. *Women, Culture, and Politics.* New York: Random House.

Davis, Angela Y. 1999. *Blues Legacies and Black Feminism: Gertrude "Ma" Rainey, Bessie Smith and Billie Holiday.* New York: Vintage.

Davis, James. 1991. *Who Is Black? One Nation's Definition.* University Park, PA: Pennsylvania State University Press.

Dawson, Allan Charles. 2014. *In Light of Africa: Globalizing Blackness in Northeastern Brazil.* Toronto: University of Toronto Press.

De Andrade, Lelia Lomba. 2000. "Negotiating from the Inside: Constructing Racial and Ethnic Identity in Qualitative Research." *Journal of Contemporary Ethnography* 29: 268–290.

de Carvalho, J. J. 2007. "O confinamento racial do mundo acadêmico brasileiro." *Padê: Estudos em filosofia, raça, gênero e direitos humanos* (encerrada) 1: 31–50.

Degler, Carl. 1986. *Neither Black Nor White: Slavery and Race Relations in Brazil and the U.S.* Madison, WI: University of Wisconsin Press.

Devereux, Peter. 2008. "International Volunteering for Development and Sustainability: Outdated Paternalism or a Radical Response to Globalisation?" *Development in Practice* 18(3): 357–370.

Dias, Diana and Maria J. Sá. 2014. "Transition to Higher Education: The Role of Initiation Practices." *Educational Research* 56: 1–12.

DIEESE (Departamento Indersindical de Estatística e Estudos Socioeconômicos). 2011. *Anuário das Mulheres Brasileiras.* São Paulo: DIEESE.

Doerr, Neriko Musha. 2013. "Do 'Global Citizens' Need the Parochial Cultural Other? Discourse of Immersion in Study Abroad and Learning-By-Doing." *Compare: A Journal of Comparative and International Education* 43(2): 224–243.

Donalson, Melvin Burke. 2007. *Hip Hop in American Cinema*. New York: Peter Lang.

Dostilio, L. D., Brackmann, S. M., Edwards, K. E., Harrison, B., Kliewer, B. W., and Clayton, P. H. 2012. *Reciprocity: Saying What We Mean and Meaning What We Say*. Ann Arbor, MI: Michigan Publishing, University of Michigan Library.

Doucet, Andrea. 2008. "'From Her Side of the Gossamer Wall(s)': Reflexivity and Relational Knowing." *Qualitative Sociology* 31: 73–87.

Douglass, Frederick and Michael Meyer. 1984. *Frederick Douglass: The Narrative and Selected Writings*. New York: Modern Library.

Downey, Gregory. 2005. *Learning Capoeira: Lessons in Cunning from an Afro-Brazilian Art*. New York: Oxford University Press.

Du Bois, William W. B. 2007. *The Philadelphia Negro*. University Park, PA: The Pennsylvania State University Press.

Du Bois, W. E. B. 1903 [1996]. *The Souls of Black Folk*. Chicago, IL: A.C. McClurg and Co.; Cambridge, MA: University Press John Wilson and Son.

Dwyer, Sonya and Jennifer Buckle. 2009. "The Space Between: On Being an Insider-Outsider in Qualitative Research." *International Journal of Qualitative Methods* 8: 55–63.

Eglin, Peter. 2013. *Intellectual Citizenship and the Problem of Incarnation*. Lanham, MD: University Press of America.

England, Kim V. L. 1994. "Getting Personal: Reflexivity, Positionality, and Feminist Research." *The Professional Geographer* 46(1): 80–89.

Epprecht, Marc. 2004. "Work-Study Abroad Courses in International Development Studies: Some Ethical and Pedagogical Issues." *Canadian Journal of Development Studies* 25(4): 687–706.

Essinger, Silvio. 2005. *Batidão: Uma História do Funk*. Rio de Janeiro: Editora Record.

Evans, Sara M. and Harry C. Boyte, eds. 1985. *Free Spaces: The Sources of Democratic Change in America*. New York: Harper.

Evans, Stephanie. 2009. "African American Women Scholars and International Research: Dr. Anna Julia Cooper's Legacy of Study Abroad." *Frontiers* 18: 77–100.

Família. 2012. "Articulação de Mulheres Brasileiras quer Luiza Bairros a frenté da Seppir." Brasília: Famílias Populares e Tradicionais. Retrieved February 24, 2013. http://www.familia.com.Br/?p=10778

Fanon, Frantz. 1952 [2008]. *Black Skin, White Masks*. New York: Grove Press.

Faria, Carlos V. 2010. "Industrialização e Urbanização em Portugal: que relações?" *Malha Urbana—Revista Lusófona de Urbanismo* 9: 79–101.

Feagin, Joe. 2000. *Racist America: Roots, Current Realities, and Future Reparations*. New York: Routledge.

Feliciano, Cynthia, Belinda Robnett, and Golnaz Komaie. 2009. "Gendered Racial Exclusion among White Internet Daters." *Social Science Research* 38(1): 39–54.

Feres Jr., João and Jonas Zoninsein. 2006. *Ação afirmativa e universidade: Experiências Nacionais Comparadas*. Brasília: UNB.

Fernandes, Sujatha. 2011. "Translating Hybrid Cultures: Quandries of an Indian-Australian Ethnographer in Cuba." In *Black Subjects in Africa and Its Diasporas: Race and Gender in Research and Writing* edited by Benjamin Talkton and Quincy T. Mills, 45–54. New York: Palgrave Macmillan.

Fernandez, Nadine T. 1996. "The Color of Love." *Latin American Perspectives* 23: 99–117.

Ferreira, Manuela and Fernanda Moutinho. 2007. "Histórias de praxe, fragmentos da vida associativa e da sociabilidade estudantis." *Educação, Sociedade & Cultura* 24: 163–192.

Figueiredo, Angela. 2002. "Cabelo, Cabeleira, Cabeluda e Descabelada: Identidade, Consumo e Manipulação da Aparência entre os Negros Brasileiros." Paper presented at the *Annual Meeting of the National Association of Post-Graduation and Research on Social Sciences – ANPOCS*, Caxambu, Minas Gerais, October.

Figueiredo, Angela and Grosfoguel, Ramón. 2007. "Por que não Guerreiro Ramos? Novos desafios a serem enfrentados pelas universidades públicas brasileiras." *Cienc. Cult., São Paulo* (59): 2. Available at http://cienciaecultura.bvs.br/scielo.php . . . Accessed on June 7, 2015.

Floyd, Samuel A. 1995. *The Power of Black Music: Interpreting Its History from Africa to the United States*. New York: Oxford University Press.

Fontaine, Pierre-Michel. 1981. "Transnational Relations and Racial Mobilization: Emerging Black Movements in Brazil." In *Ethnic Identities in a Transnational World*, edited by John F. Stack, 141–162. Westport, CT: Greenwood Press.

Fontaine, Pierre-Michel, ed. 1985. *Race, Class, and Power in Brazil*. Los Angeles, CA: Center for Afro American Studies, University of California Los Angeles.

Frankenberg, Ruth. 1997. *Displacing Whiteness: Essays in Social and Cultural Criticism*. Durham, NC: Duke University Press.

Freire, Aluízio. 2007. "Doméstica quer que agressores paguem pela violência." *Globo*. June 26. Online Edition. http://g1.globo.com/Noticias/Rio/0,,MUL59114-5606,00-DOMESTICA+QUER+QUE+AGRESSORES+PAGUEM+PELA+VI OLENCIA.html

Freire, André. 2007. "Minority Representation in Portuguese Democracy." *Portuguese Journal of Social Science* 6: 193–211.

French, John. 2000. "The Missteps of Anti-Imperialist Reasoning. Bourdieu, Wacquant and Hanchard's Orpheus and Power." *Theory, Society, and Culture* 17(1): 107–128.

Freyre, Gilberto. [1933] 1986. *The Masters and the Slaves: A Study in the Development of Brazilian Civilization*. Berkeley, CA: University of California Press.

Freyre, Gilberto. 1944 [1933]. *The Masters and the Slaves: A Study of Development of Brazilian Civilization*. New York: Alfred E. Knopf.

Freyre, Gilberto. 1998. *Casa-Grande & Senzala*. Rio de Janeiro: Editora Record.

Freyre, Gilberto. 2003 [1933]. *Casa-grande & Senzala: formação da família brasileira sob o regime de economia patriarcal*. Lisboa: Livros do Brasil.

Frisch, Michael. 1990. *A Shared Authority: Essays on the Craft and Meaning of Oral and Public History*. Albany, NY: SUNY Press.

Frustrated. 2011. Dir. Al Greeze. Al Greeze Production. Documentary Film.

Funderburg, Lise. 2013. "The Changing Face of America." *National Geographic*. http://ngm.nationalgeographic.com/2013/10/changing-faces/funderburg-text

Giddings, Paula. 1984. *When and Where I Enter: The Impact of Black Women on Race and Sex in America*. New York: Morrow.

Gillam, Reighan. 2013. "(En)countering Exceptionalism: Afro-Brazilians Responses to the Rise of Obama in São Paulo, Brazil." *Latin American and Caribbean Ethnic Studies* 8(3): 323–335.

Gilliam, A. 1992. "From Roxbury to Rio—And Back in a Hurry." In *African-American Reflections on Brazil's Racial Paradise*, edited by David Hellwig, 173–181. Philadelphia: Temple University Press.

Gilliam, Angela. 1998. "The Brazilian Mulata: Images in the Global Economy." *Race & Class* 40(1): 57–69.

Gilliam, Angela. 2001. "A Black Feminist Perspective on the Sexual Commodification of Women in the New Global Culture." In *Black Feminist Anthropology: Theory, Politics, Praxis, and Poetics*, edited by Irma McClaurin, 150–186. New Brunswick, NJ: Rutgers University Press.

Gilliam, Angela and Onik'a Gilliam. 1999. "Odyssey: Negotiating the Subjectivity of *Mulata* Identity in Brazil." *Latin American Perspectives* 26(3): 60–84.

Gilroy, Paul. 1993. *The Black Atlantic: Modernity and Double Consciousness*. Cambridge: Harvard University Press.

Gilroy, Paul. 1995. *The Black Atlantic: Modernity and Double Consciousness*. Cambridge, MA: Harvard University Press.

Gilroy, Paul. 1996. *The Black Atlantic: Modernity and Double Consciousness*. Cambridge: Harvard University Press, Fourth printing.

Golash-Boza, Tanya. 2012. *Yo Soy Negro*. Gainsville, FL: University Press of Florida.

Goldstein, Donna M. 2003. "The Aesthetics of Domination: Class, Culture and the Lives of Domestic Workers." *Laughter Out of Place: Race, Class and Sexuality in a Rio Shantytown* edited by Donna Goldstein. Berkeley, CA: University of California Press.

Goldstein, Donna M. 2003. *Laughter Out of Place: Race, Class, Violence, and Sexuality in a Rio Shantytown*. Berkeley, CA: University of California Press.

Gomes, Daniela. 2013. "About Being an Afro Brazilian Woman Living Abroad." http://www.afroatitudes.blogspot.com/2013/12/sobre-ser-uma-mulher-negra-brasileira.html

Gomes, Nilma Lino. 2006. *Sem perder a raiz, corpo e cabelo como símbolo de identidade negra*. Belo Horizonte: Autentica.

Gonzalez, Lélia. 1982. "A Mulher Negra na Sociedade Brasileira." In *O Lugar da Mulher*, edited by Madel T. Luz, 87–104. Rio de Janeiro: Relume Dumará.

Gonzalez, Lélia. 1988. "For an Afro-Latin American Feminism." In *Confronting the Crisis in Latin America: Women Organizing for Change*, 95–101. Isis International and DAWN.

Gonzalez, Lélia and Carlos Hasenbalg. 1982. *Lugar de Negro*. Rio de Janeiro: Editora Marco Zero Limitado.

Gordon, Edmund T. and Mark Anderson. 1999. "The African Diaspora: Toward an Ethnography of Diasporic Identification." *The Journal of American Folklore* 112(445): 282–296.

Greeze, Al. 2011. Frustrated: Black American Men in Brazil. YouTube video, published on February 23, 2013 and posted by BlackMansKryptonite. Accessed online June 2, 2015 from: https://www.youtube.com/watch?v=BOjvPOBvd9A

Guimarães, Antônio S. A. 1999. *Racismo e Anti-racismo no Brasil.* São Paulo, Brazil: Editora 34.

Guimarães, Antônio S. A. 2004. "Preconceito de cor e racismo no Brasil." *Revista de Antropologia* 47: 9–43.

Guimarães, Antônio Sérgio Alfredo. 2005. "Racial Democracy." In *Imagining Brazil (Global Encounters)*, edited by Jessé Souza and Valter Sinder, 119–140. Lanham, MD: Lexington Books.

Gutiérrez y Muhs, Gabriella, Yolanda Flores Niemann, Carmen G. González, and Angela P. Harris. 2012. *Presumed Incompetent: The Intersections of Race and Class for Women in Academia.* Salt Lake City: Utah University Press.

Guy-Sheftall, Beverly, ed. 1995. *Words of Fire: An Anthology of African-American Feminist Thought.* New York: The New Press.

Hale, Charles R. 2008. *Engaging Contradictions: Theory, Politics, and Methods of Activist Scholarship.* Oakland, CA: University of California Press.

Hall, Stuart. 1994. "Cultural Identity and Diaspora." In *Colonial Discourse and Post-colonial Theory: A Reader*, edited by Patrick Williams and Laura Chrisman, 222–237. New York: Columbia University Press.

Hall, Stuart. 2006. *Identidade Cultural na Pós Modernidade*, 11th ed. Rio de Janeiro: DP&A.

Hamilton, Darrick, Arthur H. Goldsmith, and William Darity Jr. 2009. "Shedding 'Light' on Marriage: The Influence of Skin Shade on Marriage for Black Females." *Journal of Economic Behavior and Organization* 72(1): 30–50.

Hanchard, Michael George. 2000. "Racism, Eroticism, and the Paradoxes of a US Black Researcher in Brazil." *Racing Research, Researching Race: Methodological Dilemmas in Critical Race Studies*, edited by France Winddance and Jonathan W. Warren, 165–185. New York: New York University Press.

Hanchard, Michael George. 1994. *Orpheus and Power: The Movimento Negro of Rio de Janeiro and São Paulo, Brazil, 1945–1988.* Princeton, NJ: Princeton University Press.

Hanchard, Michael George. 1999a. "Black Cinderella: Race and the Public Sphere." In *Racial Politics in Contemporary Brazil*, edited by Michael Hanchard, 59–81. Durham, NC: Duke University Press.

Hanchard, Michael George, ed. 1999b. *Racial Politics in Contemporary Brazil.* Durham, NC and London: Duke University Press.

Hanchard, Michael George. 2003. "Acts of Misrecognition: Transnational Black Politics, Anti-Imperialism and the Ethnocentrisms of Pierre Bourdieu and Loic Wacquant." *Theory, Culture & Society* 20(4): 5–29.

Haraway, Donna. 1988. "Situated Knowledges: The Science Question in Feminism and the Privilege of Partial Perspective." *Feminist Studies* 14(4): 575–599.

Harley, Sharon and Rosalyn Terborg-Penn, eds. 1987. *The Black Woman: Struggles and Images.* Baltimore, MD: Black Classic Press.

Harrington, Brooke. 2003. "The Social Psychology of Access in Ethnographic Research." *Journal of Contemporary Ethnography* 34: 592–625.

Harrington, Jaira J. 2010. "An Interrogation of the 'Domestic': Domestic Work, Political Subjectivity and Maria da Penha Law in Brazil." Paper presented at the *Latin American Studies Association*, Toronto, ON.

Harrington, Jaira J. 2015. *Re-Conceptualizing Rights at the Intersection of Race, Labor and Gender through Domestic Work in Brazil.* PhD Dissertation, University of Chicago.

Harris, Thomas Allen. 2001. *É Minha Cara. That's My Face.* Brooklyn, NY: Chimpanzee Productions.

Harris-Lacewell, Melissa. 2004. *Barbershops, Bibles, and BET: Everyday Talk and Black Political Thought.* Princeton, NJ: Princeton University Press.

Hartman, Eric, Cody Morris Paris, and Brandon Blache-Cohen. 2014. "Fair Trade Learning: Ethical Standards for Community-Engaged International Volunteer Tourism." *Tourism & Hospitality Research* 14(1/2): 108–116.

Hartman, Saidiya V. 2007. *Lose Your Mother: A Journey Along the Atlantic Slave Route,* 1st ed. New York: Farrar, Straus and Giroux.

Heath, Sue. 2007. "Widening the Gap: Pre-university Gap Years and the 'Economy of Experience.'" *British Journal of Sociology of Education* 28(1): 89–103.

Hellwig, David J. 1992. *African-American Reflections on Brazil's Racial Paradise.* Philadelphia: Temple University Press.

Hendrix, Katherine Grace. 2002. "'Did Being Black Introduce Bias Into Your Study?': Attempting to Mute the Race-Related Research of Black Scholars." *Howard Journal of Communication* 13(2): 153–171.

Henery, Celeste. 2011. "Where They Walk: What Aging Black Women's Geographies Tell of Race, Gender, Space, and Social Transformation in Brazil." *Cultural Dynamics* 23: 85–106.

Herring, Cedric, Verna Keith, and Hayward Derrick Horton, eds. 2004. *Skin Deep: How Race and Complexion Matter in the "Color-Blind" Era.* Urbana, IL: University of Illinois Press.

Hill, Mark. 2000. "Color Differences in the Socioeconomic Status of African American Men: Results of a Longitudinal Study." *Social Forces* 78: 1437–1460.

Hine, Darlene Clark. 1990. *Black Women in America: From Colonial Times Through the Nineteenth Century,* 4 Vols. Brooklyn, NY: Carlson Publishing.

Hine, Darlene Clark and Wilma King, eds. 1995. *"We Specialize in the Wholly Impossible": A Reader in Black Women's History.* Brooklyn: Carlson Publishing.

Hite, Amy and Jocelyn S. Viterna. 2005. "Gendering Class in Latin America: How Women Effect and Change in the Class Structure." *Latin American Research Review* 40(2): 50–82.

Holt, Thomas. 1990. "The Political Uses of Alienation: W. E. B. Dubois on Politics, Race, and Culture, 1903–1940." *American Quarterly* 42(2): 301–323.

hooks, bell. 1981. *Ain't I A Woman: Black Women and Feminism.* Boston, MA: South End Press.

hooks, bell. 1984. *Feminist Theory from Margin to Center.* Boston, MA: South End Press.

Hordge-Freeman, Elizabeth. 2013. "What's Love Got to Do With It?: Racial Features, Stigma and Socialization in Afro-Brazilian Families." *Journal of Ethnic and Racial Studies* 36(10): 1507–1523.

Hordge-Freeman, Elizabeth. 2015a. "Out of Bounds?: Negotiating Researcher Positionality in Brazil." In *Bridging Scholarship and Activism: Reflections from the Frontlines of Collaborative Research,* edited by Bernd Reiter and Ulrich Oslender, 123–133. Lansing, MI: Michigan State University Press.

Hordge-Freeman, Elizabeth. 2015b. *The Color of Love: Racial Features, Stigma, and Socialization in Black Brazilian Families.* Austin, TX: The University of Texas.

Hordge-Freeman, Elizabeth, Sarah Mayorga, and Eduardo Bonilla-Silva. 2011. "Exposing Whiteness Because We Are Free: Emancipation Methodological Practice in Becoming Empowered Sociologists of Color." In *Rethinking Race & Objectivity in Research Methods*, edited by John Stanfield, 95–121. Walnut Creek, CA: Left Coast Press.

Horowitz, R. 1986. "Remaining an Outsider: Membership as a Threat to Research Rapport." *Urban Life* 14: 409–430.

Htun, Mala. 2004. "From 'Racial Democracy' to Affirmative Action: Changing State Policy on Race in Brazil." *Latin American Research Review* 39(1): 60–89.

Hull, Gloria T. and Barbara Smith. 1982. "The Politics of Black Womens Studies." In *All the Women Are White, All the Blacks Are Men, But Some of Us Are Brave: Black Women's Studies*, edited by Gloria T. Hull, Patricia Bell Scott, and Barbara Smith, xvii–xxxiv. Old Westbury, NY: The Feminist Press.

Hull, Gloria T., Patricia Bell Scott, and Barbara Smith, eds. 1982. *All the Women Are White, All the Blacks Are Men, But Some of Us Are Brave: Black Women's Studies.* Old Westbury, NY: The Feminist Press.

Hunter, M. Gordon. 2006. "Experiences Conducting Cross-Cultural Research." *Journal of Global Information Management* 14: 75–89.

Hunter, Margaret. 2005. *Race, Gender, and the Politics of Skin Tone.* New York: Routledge.

Ickes, Scott. 2013. *African-Brazilian Culture and Regional Identity in Bahia, Brazil.* Gainsville, FL: University Press of Florida.

Instituto AMMA Psique e Negritude. 2007. *Manual de Identificação e Abordagem do Racismo Institucional.* São Paulo: PNUD Brasil.

Instituto Brasileiro de Geografia e Estatística (IBGE). 2010. *Brazilian Census.* Rio de Janeiro, Brazil.

Iton, Richard. 2008. *In Search of the Black Fantastic: Politics and Popular Culture in the Post-Civil Rights Era.* Oxford and New York: Oxford University Press.

Jacobs-Huey, Lanita. 2002. "The Natives Are Gazing and Talking Back: Reviewing the Problematics of Positionality, Voice, and Accountability among 'Native' Anthropologists." *American Anthropologist* 104(3): 791–804.

Jakubiak, Cori. 2012. "'English for the Global': Discourses in/of English-Language Voluntourism." *International Journal of Qualitative Studies in Education (QSE)* 25(4): 435–451.

Jones, Jacqueline. 1985. *Labor of Love Labor of Sorrow: Black Women, Work and the Family, From Slavery to the Present.* New York: Random House.

Joseph, Tiffany. 2015. *Race on the Move: Brazilian Migrants and the Global Reconstruction of Race.* Palo Alto, CA: Stanford University Press.

Kelley, Robin D. G. 2002. *Freedom Dreams: The Black Radical Imagination.* Boston, MA: Beacon Press.

Kelley, Robin D. G. 2015. "Baltimore and the Language of Change." *Los Angeles Times.* http://www.latimes.com/opinion/op-ed/la-oe-0504-kelley-baltimore-rebellion-20150504-story.html

Kennedy, Randall. 2003. *Interracial Intimacies: Sex, Marriage, Identity, and Adoption.* New York: Pantheon.

Kopkin, Nolan. n.d. "Does Discrimination Affect Black Entrepreneurship?" Unpublished manuscript.

Kraay, Hendrik. 1998. *Afro Brazilian Culture and Politics: Bahia, 1790s–1990s.* New York: Routledge.

Kretzmann, John P. and John L. McKnight. 1993. *Building Communities from the Inside Out: A Path Toward Finding and Mobilizing a Community's Assets.* Chicago, IL: ACTA Publications.

Latifah, Queen. 1993. *Black Reign.* Motown Record.

LaVeist, Thomas A. 2005. "Disentangling Race and Socioeconomic Status: A Key to Understanding Health Inequalities." *Journal of Urban Health: Bulletin of the New York Academy of Medicine* 82(2): iii26–iii34.

Lee, Spike. 1989. *Do The Right Thing.* Universal Studios.

Lee, Spike. 1992. *Malcolm X.* 40 Acres and a Mule Filmworks.

Lima, Ari. 2001. "A legitimação do intelectual negro no meio acadêmico brasileiro: negação de inferioridade, confronto ou assimilação intelectual." *Estudos Afro-Ásia* 25–26: 281–312.

Looser, Tom. 2012. "The Global University, Area Studies, and The World Citizen: Neoliberal Geography's Redistribution of the 'World'." *Cultural Anthropology* 27(1): 97–117.

Lorde, Audre. 1984. *Sister Outsider: Essays and Speeches.* Crossing Press Feminist Series. Trumansburg, NY: Crossing Press.

Luke, Carmen and Jennifer Gore, eds. 1992. *Feminisms and Critical Pedagogy.* New York: Routledge.

Lyons, Kevin, Joanne Hanley, Stephen Wearing, and John Neil. 2012. "Gap Year Volunteer Tourism. Myths of Global Citizenship?" *Annals of Tourism Research* 39: 361–378.

Machado, Fernando L. 2000. "Os novos nomes do racismo: especificação ou inflação conceptual?" *Sociologia, Problemas e Práticas* 33: 9–44.

Machado, Igor J. R. 2006. "Imigração em Portugal." *Estudos Avançados* 20: 119–135.

Malheiros, Jorge M. and Francisco Vala. 2004. "Migration and City Change: The Lisbon Metropolis at the Turn of the Twentieth Century." *Journal of Ethnic and Migration Studies* 30: 1065–1086.

Mancini, Jay A., Gary L. Bowen, and James A. Martin. 2005. "Community Social Organization: A Conceptual Linchpin in Examining Families in the Context of Communities." *Family Relations* 54: 570–582.

Marques, João F. 2007. "Racismo na sociedade portuguesa contemporânea: 'flagrante' ou 'subtil'?. Actas do I Congresso Internacional." Imigração em Portugal e na União Europeia. Faculdade de Economia da Universidade do Algarve. Available at: http://sapientia.ualg.pt/bitstream/10400.1/4287/1/Marques%20Racismo%20Flagrante%20ou%20Subtil%20AGIR.pdf

Martins, Pedro. 2012. "Ethnic Humour: What do Portuguese People Laugh at?" *Folklore: Electronic Journal of Folklore* 50: 87–98.

Martins, Sérgio Da Silva, Carlos Alberto Medeiros, and Elisa Larkin Nascimento, 2004. "Paving Paradise: The Road from 'Racial Democracy' to Affirmative Action in Brazil." *Journal of Black Studies* 34(6): 787–816.

Marx, Anthony W. 1998. *Making Race and Nation: A Comparison of South Africa, the United States, and Brazil.* Cambridge: Cambridge University Press.

Massey, Douglas S. and Nancy A. Denton. 1993. *American Apartheid: Segregation and the Making of the Underclass.* Cambridge: Harvard University Press.

Matory, J. Lorand. 1999. "The English Professor of Brazil: On Diasporic Roots of the Yoruba Nation." *Society for Comparative Study of Society and History* 41(1): 72–103.

Matory, J. Lorand. 2005. *Black Atlantic Religion: Tradition, Transnationalism, and Matriarchy in the Afro-Brazilian Candomblé.* Princeton, NJ: Princeton University Press.

Matory, J. Lorand. 2006. "The 'New World' Surrounds an Ocean: Theorizing the Live Dialogue between African and African American Cultures." In *Afro-Atlantic Dialogues: Anthropology in the Diaspora*, edited by Kevin Yelvington, 151–192. Santa Fe, NM: School of American Research Press.

Mazzei, Julie and Erin O'Brien. 2009. "You Got It, So When Do You Flaunt It?: Building Rapport, Intersectionality, and the Strategic Deployment of Gender in the Field." *Journal of Contemporary Ethnography* 38: 358–383.

McCallum, Cecilia. 2005. "Racialized Bodies, Naturalized Classes: Moving through the City of Salvador da Bahia." *American Ethnologist* 32(1): 100–117.

McCallum, Cecilia. 2007. "Women Out of Place? A Microhistorical Perspective on the Black Feminist Movement in Salvador da Bahia, Brazil." *Journal of Latin American Studies* 39: 55–80.

McCann, Bryan. 2002. "Black Pau: Uncovering the History of Brazilian Soul." *Journal of Popular Music Studies* 14: 33–62.

McClaurin, Irma, ed. 2001. *Black Feminist Anthropology: Theory, Politics, Praxis, and Poetics.* New Brunswick, NJ: Rutgers University Press.

McKay, Nellie. 1988. *Critical Essays on Toni Morrison.* Boston, MA: G.K. Hall.

McLennan, Sharon. 2014. "Medical Voluntourism in Honduras: 'Helping' the Poor?" *Progress in Development Studies* 14(2): 163–179.

Meyer, Michael, ed. 1984. *Frederick Douglass: The Narrative and Selected Writings.* New York: Modern Library.

Millington, Rob and Simon C. Darnell. 2014. "Constructing and Contesting the Olympics Online: The Internet, Rio 2016 and the Politics of Brazilian Development." *International Review for the Sociology of Sport* 49(2): 190–210.

Mitchell, Gladys. 2009. "Afro-Brazilian Politicians and Campaign Strategies: A Preliminary Analysis." *Latin American Politics and Society* 51(3): 111–142.

Mitchell, Michael. 1985. "Blacks and the Abertura Democratica." In *Race, Class, and Power in Brazil*, edited by Pierre-Michele Fontaine, 95–119. Los Angeles, CA: Center for Afro-American Studies, University of California, Los Angeles.

Mitchell, Michael. 2003. "Changing Racial Attitudes in Brazil: Retrospective and Prospective Views." *National Political Science Review* 9: 35–51.

Mitchell-Walthour, Gladys. 2011. "Afro-Brazilian Black Linked Fate in Salvador and São Paulo, Brazil." *National Political Science Review* 13: 41–62.

Mohanty, Chandra Talpade. 2003. *Feminism Without Borders: Decolonizing Theory, Practicing Solidarity*. Durham, NC: Duke University Press.

Morris, Aldon. 2015. *The Scholar Denied: W. E. B. Dubois and the Birth of Modern Sociology*. Berkeley, CA: University of California Press.

Morrison, Minion K. C. 1987. *Black Political Mobilization, Leadership, Power, & Mass Behavior*. Albany, NY: State University of New York Press.

Morrison, Toni. 2007. *The Bluest Eye*. Reprint edition. New York: Vintage.

Motapanyane, J. Maki. 2010. "Insider/Outsider: A Feminist Introspective on Epistemology and Transnational Research." *Atlantis* 34(2): 96–103.

Moutinho, Laura. 2004. *Razão, "cor" e desejo*. São Paulo, Brazil: UNESP.

Mullings, Beverly. 1999. "Insider or Outsider, Both or Neither: Some Dilemmas of Interviewing in a Cross-Cultural Setting." *Geoforum* 30: 337–350.

Nascimento, Abdias do. 1989. *Brazil Mixture or Massacre: Essays in the Genocide of a Black People*. Dover, DE: The Majority Press.

Neal, Mark Anthony. 1999. *What the Music Said: Black Popular Music and Black Public Culture*. New York: Routledge.

Nogueira, Maria A., Andréia M. S. Aguiar, and Viviane C. C. Ramos. 2008. "Fronteiras Desafiadas: A Internacionalização das Experiências Escolares." *Educação e Sociedade* 29: 355–376.

Nogueira, Oracy. 1985. *Tanto preto quanto branco: estudos de relações raciais*. São Paulo, Brazil: Ta Queiroz.

Noy, Shiri and Rashawn Ray. 2012. "Graduate Students' Perceptions of Their Advisors: Is There Systematic Disadvantage in Mentorship?" *The Journal of Higher Education* 83(6): 876–914.

Ochieng, Bertha M. N. 2010. "'You Know What I Mean': The Ethical and Methodological Dilemmas and Challenges for Black Researchers Interviewing Black Families." *Qualitative Health Research* 20(12): 1725.

Okediji, M. 1999. "Returnee Collections: Transatlantic Transformations." In *Transatlantic Dialogue: Contemporary Art in and Out of Africa*, edited by Michael D Harris, 32–51. Chapel Hill, NC: Auckland Art Museum, The University of North Carolina at Chapel Hill.

Okely, J. 2007. "Fieldwork Embodied." *The Sociological Review* 55(s1): 65–79.

Oliveira, Fátima, Matilde Ribeiro, and Nilza Iraci Silva. 1995. *Cadernos Geledés 5 – a Mulher Negra na Década a busca da Autonomia*. São Paulo, Brazil: Geledés.

Oliveira, Lúcio. 2007. "Tímidos ou indisciplinados?" In *Coleção Percepções da Diferença*, edited by NEINB(USP), 1–36. São Paulo, Brazil: MEC.

Oliveira, Lúcio. 2014. "Representações Sociais de Branquitude em Salvador: Um Estudo Psicossocial Exploratório da Racialização de Pessoas Brancas." *Revista da Associação Brasileira de Pesquisadores(as) Negros(as)—ABPN* 1(6): 30–46.

Omi, Michael and Howard Winant. 1994. *Racial Formation in the United States: From the 1960s to the 1990s*. New York: Routledge.

Osuji, Chinyere. 2013. "Racial 'Boundary-Policing': Perceptions of Black-White Interracial Couples in Los Angeles and Rio de Janeiro." *DuBois Review* 10(1): 1–25.

Osuji, Chinyere. 2014. "Divergence or Convergence: Black-White Interracial Couples and White Family Reactions in the U.S. and Brazil." *Qualitative Sociology* 37: 93–115.

Painter, Nell Irvin. 1997. *Sojourner Truth: A Life, A Symbol*. New York: W. W. Norton & Company.

Painter, Nell Irvin. 2006. *Creating Black Americans: African-American History and Its Meanings, 1619 to the Present*. New York: Oxford University Press.

Paixão, Marcelo. 2004. "Waiting for the Sun: An Account of the (Precarious) Social Situation of the African Descendant Population in Contemporary Brazil." *Journal of Black Studies* 34(6): 743–765.

Pardue, Derek. 2004. "Putting Mano to Music: The Mediation of Race in Brazilian Rap." *Ethnomusicology Forum* 13 (November): 253–286.

Parisi, Laura and Lynn Thornton. 2012. "Connecting the Local with the Global: Transnational Feminism and Civic Engagement." *Feminist Teacher* 22(3): 214–232.

Paschel, Tianna. 2009. "Re-Africanization and the Cultural Politics of Bahianidade." *Souls* 11(4): 423–440.

Paschel, Tianna and Mark Sawyer. 2008. "Contesting Politics as Usual: Black Social Movements, Globalization, and Race Policy in Latin America." *Souls* 10(3): 197–214.

Patterson, Tiffany Ruby and Robin D. G. Kelley. 2000. "Unfinished Migrations: Reflections on The African Diaspora and the Making of the Modern World." *African Studies Review* 43(1) Special Issue on the Diaspora: 11–45.

Peña, Yesilernis, Jim Sidanius, and Mark Sawyer. 2004. "Racial Democracy in the Americas: A Latin and U.S. Comparison." *Journal of Cross-Cultural Psychology* 35(6): 749–762.

Penha-Lopes, Vânia. 2013. *Pioneiros: Cotistas na Universidade Brasileira*. São Paulo, Brazil: Paco Editorial.

Pereira, A. A. 2013. *O mundo negro–Relações raciais e a constituição do movimento negro contemporâneo no Brasil*. Rio de Janeiro: Pallas Editora, FAPERJ.

Perry, Keisha-Khan Y. 2004. "The Roots of Black Resistance: Race, Gender and the Struggle for Urban Land Rights in Salvador, Bahia, Brazil." *Social Identities* 6: 811–831.

Perry, Keisha-Khan Y. 2008. "Politics Is Uma Coisinha de Mulher (a Woman's Thing): Black Leadership in Neighborhood Movements in Brazil." In *Latin American Social Movements in the Twenty-First Century: Resistance, Power, and Democracy*, edited by Richard Stahler-Sholk et al., 197–215. Lanham, MD: Rowman and Littlefield.

Perry, Keisha-Khan Y. 2013. *Black Women against the Land Grab: The Fight for Racial Justice in Brazil*. Minneapolis, MN and London: University of Minnesota Press.

Perry, Marc D. 2008. "Global Black Self-Fashioning: Hip Hop as Diasporic Space." *Identities* 15(6): 635–664.

Petruccelli, José Luis. 2001. "Seletividade por Cor e Escolhas Conjugais no Brasil dos 90." *Estudos Afro-Asiaticos* 23: 29–54.

Pierre, Jemima. 2013. *The Predicament of Blackness: Postcolonial Ghana and the Politics of Race*. Chicago, IL: University of Chicago Press.

Pinho, Patricia de Santana. 2008. "African American Roots Tourism in Brazil." *Latin American Perspectives* 35(3): 70–78.

Pinho, Patricia de Santana. 2009. "White but Not Quite: Tomes and Overtones of Whiteness in Brazil." *Small Axe* 29: 39–56.

Pinho, Patricia de Santana. 2010. *Mama Africa: Reinventing Blackness in Bahia.* Durham, NC: Duke University Press.

Pinho, Patricia de Santana. 2012. "Nurturing Bantu Africanness in Brazil." In *Comparative Perspectives on Afro-Latin America*, edited by Kwame Dixon and John Burdick. Gainesville, FL: The University Press of Florida.

Pinho, Patricia and Elizabeth Silva. 2010. "Domestic Relations in Brazil: Legacies and Horizons." *Latin American Research Review* 45(2): 90–112.

Póvoa Neto, Hélion. 1990. "A Produção de um Estigma: Nordeste e Nordestinos no Brasil." *Travessia: Revista do Migrante 07*, Centro de Estudos Migratórios.

Qian, Zhenchao, and Daniel T. Lichter. 2011. "Changing Patterns of Interracial Marriage in a Multiracial Society." *Journal of Marriage and Family* 73: 1065–1084.

Rachleff, Peter. 2004. "'Branquidade': Seu Lugar na Historiografia da Raça e da Classe nos Estados Unidos." In *Branquidade*, edited by Vron Ware. Rio de Janeiro: Garamond Universitária: 97–114.

Rahier, Jean Muteba, Percy C. Hintzen, and Felipe Smith, eds. 2010. *Global Circuits of Blackness: Interrogating the African Diaspora.* Champaign, IL: University of Illinois Press.

Ratts, Alex. 2007. *Eu Sou Atlântica: Sobre a Trajetória de Vida de Beatriz Nascimento.* São Paulo, Brazil: Imprensa Oficial do Estado de São Paulo, Instituto Kuanza.

Ratts, Alex and Flávia Rios. 2010. *Lélia Gonzalez.* São Paulo, Brazil: Selo Negro Edições.

Reed, Kenneth E. 1998. *Historically Black Colleges and Universities: Making a Comeback. New Directions for Higher Education, No. 102, Summer.* San Francisco, CA: Jossey-Bass Publishers. Available at: http://www.jessicapettitt.com/images/Merisotis.pdf

Reiter, Bernd. 2010. "Whiteness as Capital: Constructing Inclusion and Defending Privilege." In *Brazil's New Racial Politics*, edited by Bernd Reiter and Gladys L. Mitchell, 19–33. Boulder, CO: Lynne Rienner.

Reiter, Bernd. 2012. "Framing Non-Whites and Producing Second-Class Citizens in France and Portugal." *Journal of Ethnic and Migration Studies* 38: 1067–1084.

Reiter, Bernd and Gladys L. Mitchell. 2008. "Embracing Hip Hop as Their Own: Hip Hop and Black Racial Identity in Brazil." *Studies in Latin American Popular Culture* 27: 151–165.

Reiter, Bernd and Ulrich Oslender, eds. 2015. *Bridging Scholarship and Activism: Reflections from the Frontlines of Collaborative Research.* Lansing, MI: Michigan State University Press.

Rezende, Cláudia Barcellos and Lima Márcia. 2004. "Linking Gender, Class and Race in Brazil." *Social Identities* 10(6): 757–773.

Rhee, Jeong-eun. 2009. "A Porous, Morphing, and Circulatory Mode of Self-Other: Decolonizing Identity Politics by Engaging Transnational Reflexivity." *International Journal of Qualitative Studies in Education* 23(3): 331–346.

Rhodes, Peggy J. 1994. "Race-of-Interviewer Effects: A Brief Comment." *Sociology: The Journal of British Sociological Association* 28(2): 547–558.

Ribeiro, Carlos Antônio Costa, and Nelson do Valle Silva. 2009. "Cor, Educação e Casamento: Tendências da Seletividade Marital no Brasil, 1960 a 2000." *DADOS: Revista de Ciências Sociais* 52: 7–51.

Ribeiro, Matilde. 1995. "Apresentação: Dossiê Mulheres Negras." *Revista Estudos Feministas* 3(2): 434–435.

Rizvi, Fazal. 2009. "Towards Cosmopolitan Learning." *Discourse: Studies in the Cultural Politics of Education* 30(3): 253–268.

Rockquemore, Kerry and Patricia Arend. 2002. "Opting for White: Choice, Fluidity and Racial Identity Construction in Post Civil-Rights America." *Race and Society* 5: 49–64.

Rollock, Nicola. 2013. "A Political Investment: Revisiting Race and Racism in the Research Process." *Discourse: Studies in the Cultural Politics of Education* 34(4): 492–509.

Román, Miriam Jiménez and Juan Flores. 2010. *The Afro-Latin@ Reader: History and Culture in the United States.* Durham, NC: Duke University Press.

Romo, Anadélia. 2010. *Brazil's Living Museum: Race, Reform and Tradition in Bahia.* Chapel Hill, NC: University of North Carolina Press.

Roth Gordon, Jennifer. 2002. "Hip Hop Brasileiro: Brazilian Youth and Alternative Black Consciousness Movements." *Black Arts Quarterly* 7(1): 9–10.

Russell, Kathy, Midge Wilson, and Ronald E. Hall. 1992. *The Color Complex: The Politics of Skin Color among African Americans.* Nowell, MA: Anchor Press.

Sansone, Livio. 2003. *Blackness Without Ethnicity.* New York: Palgrave Macmillan.

Santos, Boaventura de S. and Naomar Almeida Filho. 2008. *A Universidade no Século XXI: Para uma Universidade Nova.* Coimbra: Almedina. http://www.boaventuradesousas-antos.pt/media/A%20Universidade%20no%20Seculo%20XXI.pdf

Santos, José R. dos, Maria F. Mendes, and Conceição P. Rego. 2012. "A imigração africana em Portugal nos últimos vinte anos: oportunidades e ameaças no mercado de trabalho." In *Anais do VII Congresso Português de Sociologia*, Universidade do Porto, Porto. Available at: http://www.aps.pt/vii_congresso/papers/finais/PAP0881_ed.pdf

Santos, Sales Augusto. 2006. "Who Is Black in Brazil?: A Timely or a False Question." *Latin American Perspectives* 33(4): 30–48.

Santos, Sales Augusto. 2011. "The Metamorphosis of Black Movement Activists into Black Organic Intellectuals." *Latin American Perspectives* 38(3): 124.

Sardinha, João. 2009. *Immigrant Associations, Integration and Identity: Angolan, Brazilian and Eastern Europeans Communities in Portugal.* Amsterdam: Amsterdam University Press.

Saúde Negra. 2011. "Novos ares na Seppir, Luiza Bairros Chegou." Brasília: Saúde da População Negra. Retrieved September 24, 2013. http://saudenegra.blospot.com/2011'01/novos-ares-na-seppir-luiza-bairros

Schwartzman, Luisa Farah. 2007. "Does Money Whiten? Intergenerational Changes in Racial Classification in Brazil." *American Sociological Review* 72(6): 940–963.

Scheyvens, Regina. 2011. *Tourism and Poverty.* London, England: Routledge.

Scott, Anna. 1998. "It's All in the Timing: The Latest Moves, James Brown's Grooves, and the Seventies Race-Consciousness Movement in Salvador, Bahia-Brazil." In *Soul: Black Power, Politics, and Pleasure*, edited by Monique Guillory and Richard C. Green, 9–22. New York: New York University Press.

Seigel, Micol. 2005. "Beyond Compare: Historical Method After the Transnational Turn." *Radical History Review* 91: 62–90.

Seigel, Micol. 2009. *Uneven Encounters: Making Race and Nation in Brazil and the United States*. Durham, NC: Duke University Press.

Shayne, Julie, ed. 2014. *Taking Risks: Feminist Activism and Research in the Americas*. Albany, NY: SUNY Press.

Sheriff, Robin E. 2000. "Exposing Silence as Cultural Censorship: A Brazilian Case." *American Anthropologist* 1: 114–132.

Sheriff, Robin E. 2001. *Dreaming Equality: Color, Race, and Racism in Urban Brazil*. New Brunswick, NJ: Rutgers University Press.

Silva, Daniela Fernanda Gomes da. 2013a. *O som da diáspora: A influência da black music norte-americana na cena black paulistana*. Master Thesis in Cultural Studies, School of Arts, Sciences and Humanities, Universidade de São Paulo, São Paulo.

Silva, Daniela Fernanda Gomes da. 2013b. Sobre ser uma mulher negra brasileira vivendo no exterior [About to be an Afro Brazilian Woman Living Abroad]. December 18, 2013. http://afroatitudes.blogspot.com

Silva, Graziella Morães D. and Elisa P. Reis. 2011. "The Multiple Dimensions of Racial Mixture in Rio de Janeiro, Brazil: From Whitening to Brazilian Negritude." *Ethnic and Racial Studies* 35(3): 382–399.

Silva, Joselina da. 2010. "Doutoras professoras negras: o que nos dizem os indicadores oficiais." *Perspectiva* 28(1): 19–31.

Silva, Nelson do Valle. 1985. "Updating the Cost of Not Being White in Brazil." In *Race, Class, and Power in Brazil*, edited by Pierre-Michel Fontaine, 42–55. Los Angeles, CA: Center for Afro-American Studies.

Simmons, Kimberly Eison. 2001. "A Passion for Sameness: Encountering a Black Feminist Self in Fieldwork in the Dominican Republic." *Black Feminist Anthropology: Theory, Politics, Praxis, and Poetics*, edited by Irma McClaurin, 77–101. New Brunswick, NJ: Rutgers University Press.

Simpson, Amélia. 2010. *Xuxa: The Mega-marketing of Gender, Race, and Modernity*. Philadelphia, PA: Temple University Press.

Simpson, Kate. 2004. "'Doing Development': The Gap Year, Volunteer Tourists and a Popular Practice of Development." *Journal of International Development* 16(5): 681–692.

Sin, Harng Lu. 2009. "Volunteer Tourism—'Involve Me and I Will Learn'?" *Annals of Tourism Research* 36(3): 480–501.

Skidmore, T. E. 1993. "Bi-racial USA vs. Multi-racial Brazil: Is the Contrast Still Valid?" *Journal of Latin American Studies* 25(2): 373–386.

Smith, Christen A. 2013. "Putting Prostitutes in Their Place: Black Women, Social Violence, and the Brazilian Case of Sirlei Carvalho." *Latin American Perspectives* 41(1): 107–123.

Smith, Christen. 2014. "For Cláudia Silva Ferreira: Death and the Collective Black Female Body." *The Feminist Wire*. May 5. http://www.thefeministwire. com/#article/22327

Smith, Christen A. 2015. "Blackness, Citizenship, and the Transnational Vertigo of Violence in the Americas." *American Anthropologist* 117(2) (June): 384–387.

Smith, Neil. 1996. *The New Urban Frontier: Gentrification and the Revanchist City.* Florence, KY: Psychology Press.

Solís, Patricia, Marie Price, and María Adames de Newbill. 2015. "Building Collaborative Research Opportunities into Study Abroad Programs: A Case Study from Panama." *Journal of Geography in Higher Education* 39(1): 51–64.

Souza, Neusa Santos. 1983. *Tornar-se negro*, 2nd ed. Rio de Janeiro: Graal.

Stets, Jan and Peter Burke. 2000. "Identity Theory and Social Identity Theory." *Social Psychology Quarterly* 63: 224–237.

Subreenduth, Sharon and Jeong-eun Rhee. 2009. "A Porous, Morphing, and Circulatory Mode of Self-Other: Decolonizing Identity Politics by Engaging Transnational Reflexivity." *International Journal of Qualitative Studies in Education* 23(3): 331–346.

Sue, Christina A. and Tanya Golash-Boza. 2013. "'It Was Only a Joke': How Racial Humour Fuels Colour-Blind Ideologies in Mexico and Peru." *Ethnic and Racial Studies* 36: 1582–1598.

Talton, Benjamin and Quincy T. Mills, eds. 2011. *Black Subjects in Africa and Its Diasporas: Race and Gender in Research and Writing.* New York: Palgrave Macmillan.

Tang, Eric and Chunhui Ren. 2014. "Outlier: The Case of Austin's Declining African-American Population." The Institute for Urban Policy Research & Analysis. The University of Texas at Austin. http://www.utexas.edu/cola/insts/iupra/_files/pdf/Austin%20AA%20pop%20policy%20brief_FINAL.pdf

Tatum, Beverly Daniel. 1997. *Why Are All the Black Kids Sitting Together in the Cafeteria? And Other Conversations About Race.* New York: Basic Books.

Telles, Edward E. 2002. "Racial Ambiguity among the Brazilian Population." *Ethnic and Racial Studies* 25(3): 415–441.

Telles, Edward. 2003. *Racismo à brasileira: uma nova perspectiva sociológica.* Rio de Janeiro: Relume-Dumará/Fundação Ford.

Telles, Edward. 2004. *Race in Another America: The Significance of Skin Color in Brazil.* Princeton, NJ: Princeton University Press.

Telles, Edward. 2014. *Pigmentocracies: Ethnicity, Race, and Color in Latin America.* Chapel Hill, NC: The University of North Carolina Press.

Thomas, Deborah A. 2007. "Diasporic Hegemonies: Popular Culture and Transnational Blackness." *Transforming Anthropology: A Publication of the Association of Black Anthropologists* 15(1): 50–62.

Trindade, Hélgio. 2000. "Saber e poder: os dilemas da universidade brasileira." *Estudos Avançados* 40: 121–133.

Turner, Michael. 1985. "Changing Racial Attitudes of Afro-Brazilian University Students." In *Race, Class, and Power in Brazil*, edited by Pierre-Michel Fontaine, 73–94. Los Angeles, CA: U.C.L.A. Center for Afro-American Studies.

Twine, France Winddance. 1998. *Racism in a Racial Democracy: The Maintenance of White Supremacy in Brazil*. New Brunswick, NJ: Rutgers University Press.

Twine, France Windance. 2000. "Racial Ideologies and Racial Methodologies." In *Racing Research, Researching Race: Methodological Dilemmas in Critical Race Studies*, edited by France Winddance Twine and Jonathan Warren, 1–34. New York: New York University Press.

Twine, France Winddance and Jonathan Warren, eds. 2000. *Racing Research, Researching Race*. New York: New York University Press.

Vala, Jorge, Diniz Lopes, and Marcus Lima. 2008. "Black Immigrants in Portugal: Luso–Tropicalism and Prejudice." *Journal of Social Issues* 64: 287–302.

Vargas, João Costa. 2003. "The Inner City and the Favela: Transnational Black Politics." *Race and Class* 44(4): 19–40.

Vegas, Marta, Marinieves Alba, and Yvette Modestin. 2012. *Women Warriors of the Afro-Latina Diaspora*. Houston, TX: Arte Publico Press.

Vincent, Rickey. 1996. *Funk: The Music, The People, and The Rhythm of he One*. New York: St. Martin's Griffin.

Wade, T. Joel. 1996. "The Relationships between Skin Color and Self-Perceived Global, Physical and Sexual Attractiveness, and Self-Esteem for African Americans." *The Journal of Black Psychology* 22: 358–373.

Walker, Sheila S. 1990. "Everyday and Esoteric Reality in the Afro-Brazilian Candomblé." *History of Religions* 30(2): 103–128.

Ware, Vron. 2001. *Out of Whiteness: Color, Politics, and Culture*. Chicago, IL: University of Chicago Press.

Ware, Vron. 2004. "O Poder Duradouro da Branquidade: 'Um Problema a Solucionar'." In *Branquidade*, edited by Vron Ware, 7–40. Rio de Janeiro: Garamond Universitária.

Warren, Jonathan W. 2000. "Masters in the Field: White Talk, White Privilege, White Biases." *Racing Research, Researching Race: Methodological Dilemmas in Critical Race Studies*, edited by France Winddance Twine and Jonathan W. Warren, 135–164. New York: New York University Press.

Waters, Kristin and Carol B. Conaway, eds. 2007. *Black Women's Intellectual Traditions: Speaking Their Minds*. Burlington, VT: University of Vermont Press.

Wearing, Stephen. 2001. *Volunteer Tourism: Experiences That Make a Difference*. Wallingford: CABI.

Wells-Barnett, Ida B. 1997. *Southern Horrors and Other Writings: The Anti-lynching Campaign of Ida B. Wells, 1892–1900*. Bedford Series in History and Culture. Boston, MA: Bedford Books.

White, Deborah Gray. 1985. *Ar'n't I A Woman? Female Slaves in the Plantation South*. New York: Norton.

Williams, Erica L. 2005. "Mulatas and Mucamas: Black Brazilian Feminisms, Representations, and Ethnography." In *Transatlantic Feminisms: Women and Gender Studies in Africa and the Diaspora*, edited by Cheryl R. Rodriguez, Dzodzi Tsikata, and Akosua Adomako Ampofo, 103–121. Lanham, MD: Lexington Books.

Williams, Erica. 2013. *Sex Tourism in Bahia: Ambiguous Entanglements.* Urbana, IL: University of Illinois Press.

Williams, Erica L. 2014. "Feminist Tensions. Race, Sex Work, and Women's Activism in Bahia." In *Taking Risks: Feminist Activism and Research in the Americas,* edited by Julie Shayne, 215. Albany, NY: SUNY Press.

Wolf-Wendel, Lisa E. and Marti Ruel. 1999. "Developing the Whole Student: The Collegiate Ideal." *New Directions for Higher Education* 105: 35–46.

Yelvington, Kevin A. 2001. "The Anthropology of Afro-Latin America and the Caribbean: Diasporic Dimensions." *Annual Review of Anthropology* 30: 227–260.

Notes on Editors

Gladys L. Mitchell-Walthour is Visiting Assistant Professor of Public Policy and Political Economy in the Department of Africology at the University of Wisconsin Milwaukee. She was the 2013–2014 Lemann Visiting Scholar at the David Rockefeller Center for Latin American Studies at Harvard University. She holds a PhD in Political Science from the University of Chicago.

Elizabeth Hordge-Freeman is Assistant Professor of Sociology with a joint appointment in the Institute for the Study of Latin America and the Caribbean at the University of South Florida. She holds a PhD in Sociology from Duke University and is the author of *The Color of Love: Racial Features, Stigma, and Socialization in Black Brazilian Families* (2015). She is a Fulbright US grant recipient researching modern slavery and human trafficking in Brazil.

List of Contributors

Kia Lilly Caldwell is an associate professor in the African, African American, Diaspora Studies Department at the University of North Carolina at Chapel Hill.

David Covin is Professor of Government and Ethnic Studies in the Pan African Studies Program at Sacramento State University.

Reighan Gillam is a postdoctoral fellow at the Department of Afro-American and African Studies at the University of Michigan.

Daniela F. Gomes da Silva is a doctoral student in the African and African American Studies Department at the University of Texas, Austin.

Jaira J. Harrington is a PhD candidate in Political Science at the University of Chicago.

Tiffany D. Joseph is an assistant professor in the Department of Sociology at Stony Brook University.

Lúcio Oliveira is a PhD candidate in Political Science at the University of California, Los Angeles.

Chinyere Osuji is Assistant Professor of Sociology in the Sociology Anthropology and Criminal Justice Department at Rutgers University-Camden.

Mojana Vargas is a professor in the Department of International Relations at the Federal University of Paraíba.

Gabriela Watson Aurazo is a Master in Fine Arts candidate in the Department of Film and Media Arts at Temple University.

Index

academia: and activism, 150; and black women in Brazil, 20–21; and ethnic humor, 70; and hierarchies, 64–66; and myth of racial democracy, 61; and racial diversity in Brazil, 4; and racial hazing, 73; and theory of racial democracy, 61

activism: and African American culture, 104, 181; and black feminist scholarship, 23; and black women, 16, 19–20, 25n1; and hip-hop movement, 147–148; and international influence, 5; and knowledge production, 184; and myth of racial democracy, 61; and research, 11n4, 150, 185–186

aesthetics: and domestic work, 97n7; and dominant hierarchies, 9; and nonpolitical blackness, 117; and power, 93; and self-esteem, 63–64, 181

aesthetics of power, 10, 94–96, 180

affirmative action: and Afro-Brazilian researchers, 180; and black identity, 113; in Brazil, 123; and Brazilian racial inequalities, 22; Brazilian support for, 114, 117; and Brazil's racial disparities, 164; and USF in Brazil program, 113; in US universities, 186

African American community, 85, 143, 144–145, 147

African American culture: and Afro-Brazilian scholars, 10–11, 181; and hegemonic whiteness, 161; international propagation of, 144; and

music, 141, 157; and representations of blackness, 143. *See also* diasporic consciousness

African American music: and black pride, 146; and diasporic consciousness, 142, 145, 157; and diasporic engagement, 163; and English language learning, 159. *See also* black music; *specific artists*; *specific genres*

African American Policy Forum, 183

African American Reflections on Brazil's Racial Paradise (David Hellwig 1992), 3–4, 104, 105

African American researchers, 180, 181, 185–186, 187. *See also* black researchers; black scholars; *specific researchers*

African Americans: and activism, 185; and Afro-Brazilians, 150, 167; and assumptions about race, 107; Brazilian perceptions of, 99–100, 102, 106–107, 181; and international experience, 144; and perceptions of Brazil, 61–62, 104–105

African American studies, 18, 24–25, 175

African diaspora: and black feminist scholarship, 23; and black women, 93, 153; and coalition building, 48; and common issues, 167, 178; and constructedness of race, 10; empowerment of, 187; and identification processes, 103; and identity formation, 170; and notions of blackness, 113; and

in Philadelphia, 172; and racism, 62;
in United States, 149; in United States
and Brazil, 124, 165n1, 174
Seguro, António José, 71
Seigel, Micol, 36, 102
self-classification, 83, 86
Seminário 2000, 35–36
sexism, 55–56, 86, 152, 181–182, 186
sex tourism, 24, 134, 152
sexual harassment, 152
Sheriff, Robin, 102, 119
Silva, Jucy, 48
skin tone: and interracial relationships,
123; and racial classification,
82–83, 85–86, 89n3; and researcher
positionality, 77–78, 80–81
slavery: and aesthetics, 62; and African
"nations" in Brazil, 135–136;
and black women, 16, 18, 60;
and diaspora, 103; and diasporic
consciousness, 160; forgetting of,
178; and identity formation, 142;
and notions of blackness, 114–115;
in Peru, 178; and race mixing, 85,
102, 107; in United States and Brazil,
2, 32, 164; and white supremacy, 119
Smith, Barbara, 16, 17–18
Smith, Will, 168
social identities, 86
social justice, 56
social media, 184–185
social movements, 21
social whitening, 77, 83, 84–85
solidarity, 2, 8, 16, 41, 103, 179
sotôr/sotôra, 65
soul food, 149
Soul music, 104. *See also* African
American music
Souls of Black Folk (W. E. B. Du Bois
1903 [1996]), 1
South Philly, 172, 174
Souza, Neusa Santos, 170
Souza, Raquel, 36, 38, 51, 162
Spelman College-in-Brazil, 52
State University of Bahia (UNEB), 35
State University of Rio de Janeiro
(UERJ), 125

Steele, James, 29, 38
stereotypes: and African American
culture, 159; of black bodies, 115;
and black Brazilian women, 51,
152–153; and black sexuality, 132;
and institutional racism, 165n2
Steve Biko Cultural Institute.
See Instituto Cultural Steve Biko
Stewart, Maria W., 25n1
stigma: and interracial relationships,
125, 126, 127, 128; and racial
socialization, 102
study abroad: and Afro-Brazilians,
185–186; and assumptions
about origin, 107; and diasporic
engagement, 42–43; funding for,
182–183; and global engagement,
44; and political engagement, 56; and
program development, 47, 48; and
race, 73n6; and reciprocity, 57n4.
See also international exchanges; Race
and Democracy Project; USF in
Brazil program
subjectivity, 10, 55
Supremes, The, 144

telenovelas, 62–63
Telles, Edward, 123
Temple University, 171, 175, 177
tenure, 48
Terborg-Penn, Rosalyn, 18
That's My Face (Thomas Allen Harris
2001), 104–105
transitory identity, 167
translation, 24
transnational racial optic, 10,
87–88
transnational researchers: and black
feminist scholarship, 23–24; and
black gaze, 10, 105; and blackness,
114–115; and diasporic engagement, 3;
and positionality, 2, 7, 8–9, 42,
179–180; and reciprocity, 57
Truth, Sojourner, 25n1, 184
Turner, Michael, 3, 5, 11n2
Twine, France Winddance, 5, 6, 11n2,
116, 138n29

Umbanda, 135
Unified Black Movement (MNU), 28–29, 31, 35
United States: and Afro-Brazilian scholars, 163–164, 165n3, 167; and cultural exports, 170–171, 181; and diasporic engagement, 177–178; and inequality, 174; and police violence, 175–177; and post-racialism, 186–187; and poverty, 173; and theory of racial democracy, 165n1; and transnational researchers, 179–180, 182, 184–185
Universidade Federal da Bahia (UFBA), 160
University of Campinas, 180–181
University of Michigan, 186
University of São Paulo, 180
University of South Florida (USF), 41–42
University of Texas at Austin, 150–151
urban segregation, 172
USF in Brazil program, 186; and Brazilian students, 56; and coalition building, 48; and community groups, 45; and cross-institutional interactions, 52; development of, 47; and reciprocity, 51; and research reciprocity, 46; and service-learning collaboration, 49–50. See also study abroad

Vargas, Mojana, 9, 180, 181
Verger, Pierre, 57n7
victim blaming, 69
violence, 149
violence against women, 92
Voice of Black Brazilian Women, 39
voluntourism, 43–44

Washington, DC, 172
Washington, Denzel, 170
Watson, Gabriela, 181
Wells-Barnett, Ida B., 25n1, 144
Western Journal of Black Studies, The, 28
West Philly, 172
White, Deborah Gray, 18
white allies, 48
white hegemony, 160, 161
white men: and black women, 133–134; and Brazilian women, 152; and interracial relationships, 126–128, 131–132, 136–137, 138n15
whiteness: and assumptions about wealth, 118; and brownness, 102; construction of in Brazil, 161; degrees of, 101; normalization of, 71–72; as scholarly issue, 5; in United States and Brazil, 129, 163–164
white researchers, 6–7
white women, 64, 124–128, 130
Wilkerson, Margaret, 34, 35
Williams, Erica Lorraine, 118, 152
Williams, Tonya, 39, 162
women. *See* Afro-Brazilian women; black women; white women
women's studies, 9, 15, 16–19, 20–23
Wonder, Stevie, 170
World Conference against Racism (2001), 20

X, Malcolm, 170, 172
Xica da Silva, 168
Xuxa, 168, 169

Yoruba, 135

Zimmerman, George, 176
Zumbi dos Palmares College, 147

.

CPSIA information can be obtained
at www.ICGtesting.com
Printed in the USA
LVOW04*1346310116

473068LV00008B/89/P